To

w r own
reality! Thanks for all
that input,
love Bill xxx

Publish and be damned
www.pabd.com

Create your own reality

(one I made earlier)

Gill Jackman

Publish and be damned
www.pabd.com

First published in Great Britain 2004 Gill Jackman.
The moral right of Gill Jackman to be identified as the author of this work has been asserted.

Designed in London, Great Britain, by Adlibbed Limited.
Cover ©Tony Keele : antonykeele@uklinux.net
Printed and bound in the UK.

ISBN: 1-975166-00-1

For my clients with thanks for your trust, your insights and your selves. This may or may not be of use but perhaps it's the thought that counts – and to Polly, with all my love…

Foreword

Myths and fairy stories are recognised as valuable because of the nuggets of truth they contain that mean something in our own lives. The increasing vacuousness of 21st century life sends us on wild goose chases looking for meaning, yet we still find that what we are after was within us all the time.

Unfortunately, there are now more dead ends to go down than ever. Greater sophistication, wider education, more leisure and persuasive technology have given rise to a great deal of misinformation elaborated in our fast food culture into harmful rubbish. As a counsellor I hear all sorts of nonsense and I'm keen to scotch damaging rumours with the most edible, bite-sized nourishment I could muster.

The 20th century saw the end of the Piscean age and the dawn of the Age of Aquarius. But what does this mean? Huge shifts in consciousness? Visits from Mars? Sudden enlightenment? Will it tell us what will happen tomorrow if we get up at dawn and stand on our heads? If we repeat 'I am a good person' 20 times a day will it make us believe it? Of course not. Perhaps to the more down to earth among us it means simply change and a new future. Nothing miraculous about that you may say. Not so. What if life is the miracle and it just needs to be seen clearly for us to wield its magic?

If myths have always told truths about what it is to be alive, the 20th century version got its foothold through the film industry. That's why we love Maria Von Trapp, are awed by *The Lord Of The Rings* and itch at *Trainspotting*. In this book I mine some of the raw material of who and why we are from some of the great popular movies, literature and television of our time, creating psychological popcorn of the very highest quality.

But blueprints are not enough. Maps are travelled, roads rejected, rivers bridged before we know which way to go. Change involves work, discernment and reflection. We are unique individuals even though we share an enormous amount and this book addresses that. It is about what we know after a century of psychology but it is also about what we don't. It's about keeping the baby and throwing out the bathwater. Mostly it's about the mundane plodding of the tortoise that finally beats the sparkling airhead hare to win the race and experience the paradox at the very core of reality.

Contents

Part Two
The Preparation: Heroism in Practice

Part Three:
The Separation: Karma from Dogma

Part Four
The Integration: Magic Now

Appendix

Acknowledgements and References

Introduction

I love good stories and the things they reveal and so does my daughter. I have repeatedly read and watched most of the stories this book refers to. Lots of us have. Why? Because they hold certain principles and insights that are as true in fiction as they are in real life. Art often holds up a mirror in which we can see far more about ourselves than we see in everyday life. It's just a question of looking carefully. I believe that this is why some stories are so enduring or fantastically popular but it's not always easy to hold onto what helps and discard what hinders. If these stories are true, what is over the rainbow? How do we get there? How do we use them to take control of our own stories? That is the aim of this book, and as with all good stories, we have to start with seeing what is true and getting rid of what is false.

'Won't you even try and listen for the voice behind the wall?
It cries to you
Even though it only ever whispers
Part of what it knows...
It's the fingerprint, which is never made,
It's the perfume of a rose.
And it is there if you are searching but the moment must be right,
As the night is black as the day is white
Please. Please, my friend help to make me glad
Help me find the one and only thing I've never had.
What is true?'
What is True?: Sandy Denny

Perhaps we have more in common than we think. Longing is something we can all understand. We all desire to find relief from our isolated lives by sharing them with others and by connecting to what is true for us. We yearn, we celebrate, we mourn, and we feel joy and despair. We share these feelings through songs, films, books and mostly, these days, through television. Often a seed of truth or 'bottom line' - an aspect of life that makes sense, strikes us. Often we recognise the truth of something while part of us knows there is 'something wrong'.

Somewhere between our initial attempt to say something and the meaning we have made of it the sense has been 'twisted'. Take *The Simpsons_*for instance –

"It doesn't matter how you live or what you did wrong.
As long as you're on TV people will respect you."
Bart Simpson: The Simpsons

Is it true? It is certainly something Bart Simpson has noticed. As a child, Bart is not in a position to consider or decide whether this is a good thing or not! It is simply an observation he has made about the power of the celebrity society we live in. Bart and his sister Lisa are in the process of trying to resurrect Krusty the clown's career, who is feeling a failure after being prosecuted for not paying his tax.

"What about that great feeling you get from knowing you're better
than regular people? What about being an illiterate TV clown
who is still more respected than all the scientific doctors
and educators in the country put together?"
Bart Simpson: The Simpsons

As children they take this information at face value. As thinking adults we groan and recognise the horrible truth and/or become swamped and depressed by the might of the media. Does it have to be more powerful? It takes phenomenal strength to go against the flow of 'how things are', yet the alternative is often to be crushed against the might…of what? How things really are?

The ultimate television family, we are all Simpsons: good natured, believing the best of people and corporations, using the limited information and learning that we have picked up to try to liberate and help those we love. The only problem is, we've picked up a lot of information which we have often had no chance to really think about. Homer Simpson's favourite word 'doh' nearly always occurs when he realises he's been going along with something that looks right, but it isn't. He's been duped.

Like Homer, we are constantly being duped. Just walking down the street means we are bombarded into buying stuff that promises to change our lives, which we all fall for some of the time. Increasingly, bureaucracy and the demand for accountability tells us powerful stories about what's important and what isn't. What it fails to do is tell us that filling in the right box is important to the corporation or the system. It certainly doesn't ask if it's important to us... and in the process we get further from what *is* important to us and begin to lose ourselves. Life is stressful and watching TV is easy, and human beings tend to go for what's easy. Add to that the pressure of family life, and needing a bit of peace, is it surprising that Bart and Lisa use what they see going on around them to create a picture of how things really are? Just like them, we too are sitting watching TV. Like an endless mirror of itself, in this case, we are making things the way they really are!

You can't force experience onto others but you can influence how they see things. Marge Simpson knows this intuitively:

"I guess we could get more involved in Bart's activities,
but then I'd be afraid of smothering him."
Marge: The Simpsons

The way we speak about people and the way we describe them is powerful. If it's glossy and in a magazine or said by a teacher who is good at arguing or by me as a therapist it can look suspiciously like 'the truth' but really it cannot be anything other than *my* take on you, *my* story about you. As Bart says of the opinions of his superiors when he begins to behave badly:

"You're turning me into a criminal when
all I want to be is a petty thug"
Bart Simpson: The Simpsons

We have to struggle to reach our own conclusions through our own efforts, about what we do, about who we are, about who we would really like to be. Yet we (each of us) have absorbed so many ideas already. Is it any wonder that the things that unite us the most are basic human feelings and that the stories we are most fond of reflect our hopes, our dreams, our sadness which provide a connection we understand instinctively?

What is it about these stories that is true? Television is powerful. Whether it creates our experience or simply resonates with it, the stories are often so intertwined around the feelings that we cannot tell the difference between the feelings and the beliefs we have absorbed. Yet there is a difference and this difference is the key to a better system of living, a system where we can become the people we really are. It is one thing to say that I fear the influx of asylum seekers and its influence on my culture (a feeling) and quite another to say that they should all be banned because they're a bunch of workshy skivers (not only a belief but a strong prejudice). The stories on TV or in the papers get inside us, teaching us things we may not agree with but without careful examination we are learning a point of view all the same. Perhaps more relevant, the stories we have made don't simply come from the media – some of their origins are deep in our own beginnings and far more pernicious than we can imagine. Oblivious to them we accept them, believing that we are living life to the full just because they drown out everything else.

In *The Simpsons* at the Itchy and Scratchy focus group the marketing men are frustrated that kids clap whatever cartoon they're shown. The programme maker is annoyed that they don't provide him with a clear preference so he'll know what will sell.

"You kids don't know what you want. That's why you're kids. 'Cos you're stupid."
The Simpsons

A fantastically adult perspective, and because it doesn't suit him right in this moment he's cross that kids are open, naïve and gullible. He hasn't considered that we, as adults, might be responsible for what we put into their mouths, their heads and their hearts. The whole point about the *The Simpsons* is that it is a programme about *what people do* with the information they pick up – Bart, Homer, the nasty man trying to sell a TV programme, they all have their own way of making sense of the world. If we can see the stories *we* swallow in this light, *we* can make choices about what *we* do with them.

Imagine if we could be absolutely confident in how things feel *for us* and could communicate that in a way that people would find difficult to ignore or ridicule. Imagine if the inner knowing could influence the outer world even if it was just with your friends and family. The inner knowing would have to be granite where now, for a huge amount of people, it feels like putty.

This is a book about alchemy, about turning putty into granite and how to use that granite as an anchor. It offers a way to explore our beliefs and how we perceive while giving us a deeper access to ourselves. Conscious of the ways of thinking and behaving we fall into we would have a great deal more choice over the way others see us and there is every chance that we could challenge *their* perceptions and fixed beliefs. Most of all we would no longer be trapped in our own restrictive patterns. Once we got really adept at it we would literally be wielding a magic which changes reality.

To turn putty into granite we have to get hold of our root feelings and the beliefs we have developed out of them – we need understanding and courage – we need the stories of others – how they have grown and how we have grown. This book links up with our fictional big and little screen and literary heroes and heroines to get closer to the stories of our own lives. By looking at how they deal with *their* lives it aims to show what *we* do with the information *we're* given, tracking how we develop beliefs and make choices, or not, in our own lives. *Buffy the Vampire Slayer,* Kat, from *Eastenders, Harry Potter,* Gandalf and Frodo from *The Lord of the Rings,* Renton from *Trainspotting* - all these characters are massively popular for the simple reason that people have an intuitive grasp that there is something real about them, they see something of themselves in them – a seed of truth that can grow.

Most of all we need to find what is at our individual core and at the end of each chapter there is an exercise that is not for the faint-hearted. Magic is a serious business, but in my experience, true seekers are rarely faint-hearted.

How We Grow

As anyone who knows *The Simpsons* can see, Marge is a great mother and Bart and Lisa are emotionally resilient because of it. They are strong in some very necessary beliefs. If someone told them that their mother didn't love them, their response would be to walk away, convinced that the person was crazy. It is not difficult to imagine that however many versions of Rambo they saw, they would never seriously behave like Rambo, but there are subtler influences, deeper layers of belief that we don't even know are there and there are plenty of people who do not have a mother like Marge. Our beliefs start to be formed at a very early age so it is very important that this book goes right back to explaining just how vulnerable we can be. Bart and Lisa have been fortunate, they have enough love in their lives to know a horror film from true horror. For the rest of us it might be useful to start at the very beginning and check it out.

Part One
The Revelation:
How We Work

Chapter One:

What We Do

If we're secure then that is what is 'real' to us, and it seems as if the whole world is secure, but what if it isn't? What happens if something disturbs our sense of what is solid, secure or permanent before we have grown enough for it to be substantial and shockproof? What if mum disappears or is depressed or ill or violent when we are too young to understand? What happens if someone hurts us? What if dad beats mum or beats us just as we are weaving the whole fabric of our world out of what is around us? How do we see our world then? What do we do before we are able to blame others for our difficulties? Even before we have any kind of grasp of other people? What happens if we suffer, are hungry, are hurting, yearning for love and nurturing and it doesn't come?

Huge holes develop in the security blanket of our world.

This is a book about what we know about the holes in our world. It is about how our world gets torn apart at certain times and not others. It is about whether the holes can be mended and, if so, how. This book addresses what makes them bigger and what makes them smaller and what we filled the holes with when they first appeared.

Perhaps it is useful to say here, that unable to blame the people who are supposed to love us, lest they never ever love us again, we blamed ourselves, plugging the gaps with self-hatred. I wonder how much we still do?

The Internalised Plastic Surgeon

Sometimes it seems as if we are a two tribe world, where the extent to which we have been neglected, abused or nurtured determines whether we take a scalpel to our insides in an attempt to cut out our 'bad bit' or whether we try to accept all the bits and get them to talk to each other and make friends. How well entrenched our beliefs are has an impact on how sharp and solid that scalpel is.

In David Kronenborg's film *Videodrome,* the main character goes on a killing spree with a gun that is made from his own hand. It is part of his body - a symbol of violence so familiar that there is no separation between it and him. His hand *is* a gun. – a part of himself that he turns on others but could just as easily turn on himself. If we are abandoned or not looked after when it was crucially important, a violence is generated in *us*. Those of us who were forced to cope in some way, without understanding why, often believe that everything is their own fault and that these things happened to them because they were not good enough people. But where does the violence go? By the time we are adults this pain is so much part of *us* that we can't see it, even though everyday we turn it into a scalpel with which we attack ourselves and others.

Believing we are already bad, we fill our heads so full that there is no room for what we are feeling (for instance, feelings of jealousy, rage, joy, sadness) and we violently argue away our true experience with the same kind of ideas, attitudes or 'messages' as our mothers or cultures put in our heads. When Marge starts to have a nervous breakdown in *The Simpsons*, these are the commands she endlessly repeats to herself, all the time making it worse, beating herself up instead of allowing her true feelings the sympathy that they deserve.

"Be good, behave, behave, be nice, smile,
be polite, don't make waves."
The Simpsons

Having cooked cakes all night in an attempt to be 'good enough' she cannot stop for a break even when Lisa urges her to, because:

"I'm keeping myself in a state of cat-like readiness."
The Simpsons

For what? For the deep feelings of shame that are emerging. As we discover later at the psychiatrist's, avoiding the feelings by constantly proving she is 'good', in this case by baking cakes, has become such a bad habit that it is automatic. It 'drives' her behaviour. The sense of shame is constant. It rolls down well-worn grooves carved in her psyche doubtless from some long ago sense of failure. The feelings of shame then get changed into a belief which is less unbearable than raw shame (I am worthless) neatly short-circuited by the action plan and more palatable belief:

- If I try harder I will be loved in the end.

Hence the cakes.

This is not helping, but merely feeding, arming and escalating the war within with what I call the 'inner fascist'! As Marge says:

> *"Judge jury and executioner all rolled into one."*

The foremost expert on this process is a psychotherapist called Alice Miller. She defined the individual psychological stages in the lives of most people as:

1. To be hurt as a small child without anyone recognising the situation as such.
2. To fail to react to the resulting suffering with anger.
3. To show gratitude for what are supposed to be good intentions.
4. To forget your own felt experience.
5. To discharge the stored up anger onto others in adulthood or to direct it against oneself.

> *"The greatest cruelty that can be inflicted on children is to refuse to let them express their anger and suffering except at the risk of losing their parent's love and affection. The anger stemming from early childhood is stored up in the unconscious, and since it represents a healthy, vital source of energy, an equal amount of energy must be extended in order to repress it."*
> *Alice Miller:* For Your Own Good

Remoulding the Inner Self

I first met my inner fascist in a dream where I had to go to a processing plant that cleaned filthy chemicals out of beach sand. I had put my clothes on the beach earlier and they had been sucked down and away. By the time I reached the huge factory I was freezing, almost naked and deeply vulnerable. The figure in my dream resembled a female concentration camp guard. She showed not a glimmer of humanity, refusing to speak to me or acknowledge or anticipate my dilemma so that I felt like I was speaking to a brick wall. I find it helpful now to visualise that it is she who, on very bad days, orders me around, tells me I'm not good enough and fills me with voices declaring their contempt. This fascist is capable of using every argument and every piece of evidence to treat me like dirt. On occasion the blaming can be so crushing that I can no longer stand it and so hurl it out onto someone else, so instead of:

- I'm a useless, stupid, incompetent, clueless bitch/bastard.

I blame whoever is nearest and say:

- You're a useless stupid incompetent bitch/bastard.

It took years of throwing this phrase around like a squash ball before I realised that that was exactly what it was. A ball of energy that had to go somewhere, generated by the conflict. Imagine for a moment two statements:

I'm bad   I'm good

Now pretend those statements are twigs, rubbing together faster and faster as each tries to rub the other out. If this is what happens to you, you can guarantee that the inner fascist is running along behind you cracking a whip. Frequent sparks and exhaustive explosions of energy are bound to be the outcome.

The notion of finding and 'rooting out' the negative statement is now quite familiar in therapeutic circles but of course, if it is still your inner fascist doing the 'rooting' the basic problem will get worse. This kind of violence is simply the inner fascist getting his or her way. A qualified therapist would be constantly on the watch for how a client was receiving/hearing their input and what they were then doing with it which is why unqualified therapists can do such damage and why I have a problem with NLP (a technique for reprogramming the mind) and self help books that work with affirmations generally. Under the tutelage of the Inner Fascist, the optimistic phrase:

You can 'Heal Your Life' becomes:

"Come on get on with it. Everyone else has."

"I AM really pathetic, and hopeless and fat. And look at me, thinking of myself again."

"Oh you're pathetic. Always whinging on like you're the only one in the world who has problems"

"Yes, and I was just going to think, 'I must be really selfish'."

"You always have to have the most difficult pain. You're totally self-obsessesed and you just refuse to shift."

"I'm just being awkward and difficult."

"You think you're so special. Why is it always you who should have special treatment. You're manipulative, attention-seeking and a massive drain."

"Oh God, I must stop thinking about myself."

"It's my fault my life is such a mess."

"Yes I must try harder. Everyone else finds it really easy and can do it. I just feel so shit."

"You *can* heal your life, but you won't."

And here we are, full circle, back at the belief about how worthless and shit and BAD we are.

This is the point where people start to bang their head against the wall.

If you recognise this process, this book should help.

Because of the way we have developed as a culture, we have come to believe that the only way of moving forward is through finding a problem and solving it. We even come to assume that:

• Being an adult means finding a solution.

Because of the way we're made, we love to ask questions and find answers. That's our *mechanism* for growth, but the growth itself lies in standing outside the cycle of question and answer and accepting how limited that is, as well as how valuable. It's not the answer to everything but it's got *a* purpose.

Imagine a child getting up in the morning with a terrible headache, confused and covered in spots. The child knows little about illness, has no sense of its own wellbeing and clings to what it does know. Because it is a school day, the child insists on going to school. This is entirely consistent: It's a school day, therefore I go to school. From outside, however, we have privileged information. This child is ill and needs rest in bed with some care and appropriate softness.

As a counsellor I see adults do just the same thing all the time. They come in great distress and proceed to beat themselves to a pulp, demanding an answer from themselves. The greater the distress, the more viciously they treat themselves. Frequent comments are:

1. Why am I so useless?
2. Why didn't I see that?
3. How could I have been so stupid?

When I'm low on energy reserves myself I soon revert to my old strategy for coping and out trots the inner fascist, usually in a new disguise, asking:

1. Why haven't you sorted this out yet?
2. Why haven't you let go of that?
3. Look at you – you've got nowhere.

However, the inner fascist is always recognisable by the level of violence and harshness in her voice or in her commands. Vitality is not violent or harsh, and the above questions, even when answered, do not provide the necessary healing for my stress or my degree of overload. Instead of the inner fascist, what I need is the inner mother to come and soothe my fevered brow (however pathetic! Ideal mother's love is not conditional!) so it's off to bed with a cake, the paper and a film. Whatever calms me down. When I cannot cope, I need to rest, just like the child with measles needs to go to bed and consider, with awesome joy, how many questions she doesn't know the answer to.

So – we remould the inner self by *including* all the bits. It's not complicated to do this, just very difficult. I have clients who don't know the meaning of the word 'comfort' or who have been programmed to despise certain parts of themselves so deeply that reconnecting with them without judging them is unimaginable. These parts become dangerously wounded, all the more so because once the injury is done, the desire to ignore, cut out or spit on that aspect of oneself has become habitual behaviour and, as we all know, a neglected wound is likely to become gangrenous. Our real mothers knew this and attended to physical injuries. My own mother's favourite phrase was usually:

"hold still or it will go septic"

The same is true of emotional wounds. It might even be that what was a light graze, some slight or criticism that was ignored or neglected or that led to ridicule could eventually fester and lead to a high fever which could eventually weaken a part of us chronically.

<u>It's All My Fault</u>

Once we become convinced of the reality of our fixed beliefs, the notion that:

- It's all my fault becomes as familiar and inevitable as the notion
- It's all your fault.

Things seem black and white, good and bad and swing wildly between total self-blame and blaming the other. I often have clients who are focusing on how wrong they are because they don't 'fit' into their marriage, or because they don't 'trust'. The basic message 'I'm not good enough' and the feelings of shame that they are trying to avoid prevents them from seeing that their response might also be connected to the way their partner treats them. The status quo still pressurises people to conform and persecutes those who don't. Perhaps that's why so many self-help books are about *how to be* something other than what you already are. Of course it's very useful to know ways of managing your anger, but only if that doesn't involve pretending to be something you're not and ignoring what you may need desperately to express, even if only to yourself. So many people seem to want to stop their 'inappropriate responses' but whose definition are we talking about here?

As Lisa Simpson says:

'You can't be cool just by spouting a
bunch of worn out buzzwords.'
The Simpsons

We all know that Lisa is exceptionally smart and you don't have to look further than her for some very sensible advice. There is nothing solid about buzzwords but from studies and observations throughout the last century there are certain things we know for sure about what messes us up and what makes us healthy; about how we are programmed to develop and what happens if we don't get what we need. The appendix to this book contains a potted version of some of them while the book itself addresses what we often do.*

As you can see, if we believe the worst of ourselves, but can't quite see it, all the self-help books in the world are not going to help. Our response to stories tells us a lot about ourselves but we rarely wake up one morning having worked it all out. Our lives appear to be like a jigsaw puzzle. We go back to a piece, again and again, and suddenly 6 pieces fit together all at once.

If any of these stories touch you more than others, read or watch them again. Follow what *moves* you rather than trying to *get it right.* That's where the magic is. Most books start at the beginning and go to the end. Not this one. Like the more accurate notion of *cyclo*therapy (Irving Yalom) as opposed to psychotherapy, it goes round and round in circles trying to get the penny to drop, so there is no *right* way to go. After all, if it's something specific that stops us getting the benefit of all the wisdom that's around, it's worth returning to it until it makes sense! If you like, you can choose to find out more about the giant theorists behind this information by looking them up in the appendix in the back of this book. Or then again, you might want to read the outlines at another time altogether, or not at all. It really only matters to you.

* For instance, this may be a good time to look up Bowlby, depending on how you feel.

Chapter Two:

What We Need

'Perhaps I had a wicked childhood,
Perhaps I had a miserable youth,
But somewhere in my wicked, miserable past,
There must have been a moment of truth.
For here you are, standing there, loving me,
Whether or not you should
So somewhere in my youth or childhood
I must have done something good.'
Rogers and Hammerstein: The Sound of Music

Julie Andrews as the ex-nun Maria experiencing salvation in the arms of Captain Von Trapp, played by Christopher Plummer, still brings a lump to my throat. Perhaps the *moment of truth* she was referring to was the split second realisation that she was worth something; that as a child she had a moment when absolute unconditional, glowing, endless love flowed into her core and she *knew* it. She may have gone on to forget about it, to undervalue it, or more likely to hide it under layers of convent liturgy but Captain Von Trapp reminded her that it was there.

'Nothing comes from nothing,
Nothing ever could'
Rogers and Hammerstein:
The Sound of Music

Of course, we are brought up to believe that if we're good we'll be rewarded and Maria, in her song, does just this, but just stop for a moment and imagine that instead of Captain Von Trapp being her reward for being good, it is simply that Maria can accept his love because deep down *she knows she is really alright.* What if Captain Von Trapp, instead of providing the love, *reminded her* of the glory that was her heritage; reminded her of what it was to be loved. Where did it come from, this glory? If you remember, Maria was always getting into trouble simply because her joy in life meant she was whizzing around on top of the Alps instead of praying. My guess is that the fictitious Maria didn't have too much of a wicked childhood at all, but had a *good enough* mother.* With clean air, healthy living and additive-free food, the tiny spark that was her life force grew until her cup overflowed.

Grounding

Grounded is a term so bandied about in the new age movement that its meaning is rarely looked at seriously, yet how grounded we are is a good and measurable indication of work we may need to do to become more complete. The term *down to earth* is probably the nearest in everyday use that we get to the phrase *being grounded* but it can be used as a mask that proclaims 'I know what's what' when nothing could be further than the truth. Behind every *overly* down to earth person there's usually a very feminine type who's terrified she might float off and not come back!

Let's return to our heroine. When I think of *The Sound of Music* I think of seven children, dressed in brightly coloured curtain material, swinging from the trees. When Captain Von Trapp comes home he is furious at this chaos since he usually runs his home with fear and a whistle. He is probably so terrified of over-indulging his kids that he falls back on what he knows: the discipline of the navy. Maria has no such fear or qualms. She is sure enough in her beliefs about what is right and what she can manage that she can bend the rules occasionally without the family falling into anarchy. She looks like one of those daffy, quirky types but by the time you see her bringing up the rear, holding the hands of some of the youngest children as she strides across the isolated Alps to Switzerland, you know that her core is made of solid steel.

She is thoroughly *grounded.*

She knows exactly what is important to her. As a consequence of this she is *authentic* in her action. When the Nazis come, there is no prevarication. Nobody says, 'perhaps it will be alright' or well 'we're not Jewish so what does it matter to us?' or 'life as a big guy in the navy will be pleasant enough'.

Why not? Well, perhaps in this fictitious film with its very clear convent

* This is Winnicott's term (appendix) and is very useful in helping us manage our perfectionism.

rules we could put Maria's groundedness down to an upbringing which taught right from wrong at an early age, but it's clear that there is more to it. At the beginning of the film when she's late (again) she questions whether it can be wrong to be late if she has been delayed by wonder at the beauty of God's earth. A very good question! So, she's a free thinker as well? If her confidence came from the convent it would rely entirely on her following the rules *exactly.* Veering away from them would be terrifying but for her it's no big deal. She's not even rebellious. She is *authentic** and she knows what she feels and believes so she follows her heart.

So easy to describe. So difficult to be for those of us who have been damaged. I'm sure there are many people who can't stand *The Sound of Music*, but I think a lot of this is envy! Some of the people who struggle the most are trapped in a loop where they are trying to answer the question:

"How do you know how to be/what to choose/who you are?
Are you it already or do you just make a decision?"

The problem that they face is an inability to describe or to fully experience their feelings. There are many reasons for this as well as solutions, often to do with not recognising the importance of our own experience and so not giving it any attention. Being grounded is typically a chicken and egg situation: you can't be grounded unless you feel and recognise your feelings; you can't feel and recognise your feelings until you are grounded. But there is always some sense of oneself to work with.

So what does it really mean? To be grounded is to:

- Be able to state your beliefs and what is important to you
- Be able to change your mind on reflection
- Be able to say how you feel at any given time.
- Be able to say that you're unsure what you feel right now, but you'll think about it!
- Be able to give a realistic account of your strengths and weaknesses,
- Be confident that you have the right to prevent people taking advantage of you
- Be able to differ from other views while accepting their validity

*A term used in counselling to refer to ones *own* felt experience whatever others may think or feel.

So what are we talking about here? An assertiveness course? A determination to become more broadminded? No technique or answer to the philosophical question above can teach you to be grounded. *Know thyself* is the only way because it is only possible to be authentic when this is the aim. Techniques such as thinking, shouting, praying might all help, but only because they help us to know who we are, which leads me back to Maria's steel core. It's alright for her but what are we supposed to do if we haven't got it?

Envy might help! Why? Because, much as you may dislike it, it may be your most accessible authentic feeling. It's probably even justified, but my guess is that you turn from it and deny it. Who wouldn't? It's horrible. However, it might be all you've got on offer right now though there are bound to be other feelings, most likely out of reach. You might be flying into a rage right now reading this. I apologise for that and hope that this book at least gets you closer to deciding the way forward. There is an exercise on grounding at the end of this chapter, so you can simply cut to the chase and do it right now if it's most helpful, *because what grounds us the most is our pain.*

At no point in *The Sound of Music* does Maria have a problem expressing what she feels. She is not afraid of her feelings and none of her feelings are unacceptable. Even her jealousy falls into place and helps her to realise that she is in love with the Captain, who is preparing to marry someone else and therefore it is not appropriate to stay in his house. This is because she is able to accept and acknowledge her feelings as part of her, so when they knock on the door of her being she does something fundamentally different to what damaged people do.

• *They* tell themselves that these feelings are not OK and shove them away either with a little voice such as: 'Oh don't be pathetic, you knew this would happen', or with a little voice saying: 'what's the point of feeling that. I'll only get upset and it won't change anything.'

In contrast, Maria opens the door saying: 'oh dear, you *are* upset, what's the matter' and then listens to the answer.

This is not even something she has to make an effort to do (more reason for envy). It's all very straightforward for her as she has obviously never been seriously thwarted in her feelings or abused. For us, turning around our way of relating can take forever, and if you want it to become authentic you have to ground it in your experience – and that means fully feeling whatever is happening, with no judgements and no actions. You don't have to do anything. You just have to be. There is no solution. *By being allowed our emotions, our traumas and our struggles, however ridiculous they seem, we get to really grow up and to leave childish things behind.*

Permission

So who will allow us? As human beings our highest ideal is the freedom which we consider our birthright, yet we constantly censor what we feel and what we do, often not from choice but out of fear:

- Fear of what we might do to others
- Fear of what others might do to us
- Fear of what we might discover
- Fear of being criticised
- Fear of being disapproved of

We do most of it to ourselves to prevent being hurt, especially the last two. We disapprove of our own feelings using the same messages we learned as children:

- There's no point in getting upset
- You're only doing it for attention
- Pull yourself together
- Big boys don't cry

We actually say these things to ourselves when we start to feel the familiar lump or the rising anger. Is it any wonder that this often exaggerates and feeds the distress? And we've been doing this since we were children. Our feelings are stuffed so far down into the deep dark lake it's not surprising that we're afraid, if we look too hard, of what we'll find. Maria never has this problem but then her feelings were never denied the light of day and, consequently, they never really damaged her or anyone else. They hurt, sure, but what are a few tears now and then?

So we need permission instead of denial, and as adults we have the freedom and the will to actually do this for ourselves. If you can imagine yourself as both an adult and a child and get hold of the voices of each you can separate them, recognise them and use this as a tool to stop attacking yourself and let the emotions flow. Later in this book there are exercises to facilitate this.

Benign Mirrors

How many fairy tales there are with mirrors in them! In *Snow White* the Queen looks in the mirror to boost her confidence. In this case it is important to her to be told she is beautiful. The *most* beautiful, but she might as well be asking: 'Am I alright?'

We all have doubts, and when we do we seek reassurance. We look in the face of our partners, as we did our parents in search of evidence that we are alright. When we have particularly strong inner fears that we are *not* alright, then that is what we look for and that, generally, is what we find, imaginary or otherwise.

I refer of course to what we do as adults. How much we do it depends on how fully formed we are. If we are inauthentic and insecure we will take more of a cue from what we perceive, in the faces of others, to be a reflection of ourselves. If we are authentic and know ourselves, our strengths and weaknesses, we see the other, and while appreciating their approval, our wellbeing and sense of who we are is not so dependent on it. This is why two people who are insecure can spiral downwards in a relationship at such high speed.

As children we are not fully formed. That is certain. While debates continue to flourish about whether we are born a certain way or made that way by the environment, it is agreed by most that it's a bit of both. That is, what can be spoiled by nature can be helped by nurture and what can be helped by nature can be spoiled by nurture. Our parents and carers act as mirrors as we are forming. When I look in the mirror now, prior to leaving the house or if seeing a client, I check to see if 'I'm alright'. Have I got crumbs on my cheek or mascara smeared beneath my eye? Every time our children turn to us before running off to play or after throwing something across the room, they are asking: 'Am I alright?' Being children, our children, who we love in a way that no one else does, we usually think they are the bee's knees, even if occasionally their behaviour is unacceptable. So we effectively hold up a mirror. We approve of all that is acceptable and disapprove of what is unacceptable. What we are doing here is creating their *self-image.* Did someone say bringing up kids is a huge responsibility? It is.

Benevolent Authority

Parent's fear of damaging their children has a lot to answer for. The authoritarian post-war years spawned a lot of idealism, people who thought that if you let the kids run free they would grow up to be beautiful people. There was a fear that putting your foot down was putting the boot in, mainly because when their parents had done it, it was.

It isn't. Imagine your face as a child in the mirror. Is it good or is it evil? If you were hated, maybe with those repressed emotions we talked about earlier, you would probably think evil lurked behind it. However, if you were chastised for using all the washing powder to wash your dolls clothes, perhaps sent to bed early, not given sweets or not allowed to watch your favourite TV programme, but were still kissed goodnight and obviously loved, you would not. You would look in the mirror and say: 'Ah, I'm ok and I shouldn't have done that.' You might even say: 'I hate my mother, she's so unreasonable' and throw a tantrum, but still you would learn that these strong and violent feelings pass, the house doesn't fall down and you are *still* loveable. Better and better.

There is nothing more alarming to me than seeing parents giving in to children that throw tantrums in supermarkets because they want some toxic sugary rubbish. If the child's higher self could speak it would say:

'*Speak to me gently, even if I shout. Offer some fruit. If I continue to scream put up with it as cheerfully and as best you can. Acknowledge what I want but hold to your guns and explain why. You are teaching me a very valuable lesson that I will have to learn sooner or later. That disappointment in life is tolerable, I can live through it and my emotions are perfectly safe. They neither destroy others or myself.*'

Of course the child does have a higher self. It's called *The Parent*. When it looks in the mirror next time it sees what they have taught it.

- That I am lovable,
- I am worth listening to,
- I am deserving of good things,
- I am not dangerous,
- I am manageable and loved.
- Life is sweet and quite safe.
- Authority is likely to act in my best interests.

As we grow up, if this is consistent, it becomes *internalised*: the benign authority *becomes* that authentic core so that when we feel jealous (like Julie Andrews) or are pursued by Nazis or left by husbands or wives, we fall back on the self image made up of these beliefs. The core belief 'I'm alright' means that dealing with the inevitable pain of life is relatively straightforward.

Visibility

As children we need to be seen realistically. Our shortcomings and successes need to be seen in perspective with the constant proviso of: 'I love you anyway.' Christianity is possibly the closest we come - in the UK - to a grasp of how this works. Ask any practising Christian what's so great about Christianity and they will answer: 'Jesus loves me unconditionally. No matter what I have done, no matter what my failings are, he can see them and it makes no difference.' This love, or belief in love, is the thing that then sustains them when the going gets tough and roughly translated it leaves people believing that 'I'm alright'.

If we feel guilty, or something is bothering us it is often a relief to be seen, 'warts and all'. If a close friend can truly see me and thinks I'm not so bad, I can accept myself more easily. As children, if our parents can't cope with shyness or frustration, loud noise or us not understanding something, they can become quite angry and, to us as children, frightening, so we tend to hide these qualities. If we do this for long enough they become hidden from ourselves. We may find that we despise those very qualities in others that we no longer realise we have in ourselves. I'm sure you can imagine how badly this affects our attempts to be authentic. If we become ashamed instead of aware of who we are and what we feel, is it surprising that we keep cutting off from our feelings? Feelings of shame can be the very worst.

If certain qualities are not recognised or seen by parents, or if they are severely disapproved of, we develop a *false self* of which they do approve. I, for instance, was always terribly sensible and coherent with my father because if I cried, wailed or didn't know the answer to something he would be dismissive. I know that now, but for years I had a terrible fear of bursting into tears in front of him and of course, any approval I did get from him could never satisfy me because I was left realising that if only he knew the hidden part, he would not approve at all.

Attention

Unfortunately, those trying too hard to be good tend to see their emotional eruptions as badness getting the better of them. They are in fact, a clue to what needs attention and point towards that which has been buried underground for too long. In this way there does seem to be some redeeming divine balancing mechanism in the universe. Granted, you may seem out of control, and on the surface, pretty unacceptable but any flower or plant that gets no light is going to emerge a bit deformed. What is so extraordinary about this mechanism is that it will be very persistent in trying to be seen or heard until it gets the attention it deserves. To build a sense of who we are, we need to be heard in our fullness as much as we need to be seen. With regard to the child in the supermarket 'No you can't have that,' can be mediated with an acknowledgement that 'I know you really want that'. If someone isn't heard, they will simply repeat their behaviour until they are.

Recently, I found myself in the pharmacy of my local hospital awaiting some antibiotics. There was a huge queue, and this set me in a panic. I found I was becoming louder, very demanding about the wait and fearful that I was disappearing beneath the anonymity of it all. Consequently I began to argue. The point I was making was that while there were many people awaiting long prescriptions mine was a single item, easy access, and minor medicine. I had a car on a parking meter and a daughter to get home to. I had already taken one antibiotic, in the casualty department and if I left the department without the medicine the resulting gap would make the antibiotics ineffective.

In actual fact, what I was really saying was: 'I can't see a solution to this problem and you don't care. I feel panicky already in here and you can't see me. Well you're bloody well going to see me.' I didn't need them to make me jump the queue or tell me I was better than everyone else, I just needed to be heard. I was discounted by the system and found this so threatening that I was going to go round and round saying: 'Can you see me, can you see me?' until someone recognised my difficulty and I felt *visible.* That was all. The way in which this can get quite out of hand is explored in the next chapter.

Why Counselling?

It's a difficult generalisation to make as there are so many different schools of counselling so when I speak about counselling I am referring to what is called *person-centred* counselling, of which there is a full description in the appendix under *Carl Rogers*.

When we become upset as children, if we are lucky, we have mothers who not only ask: 'What is the matter?' but frequently intuit the problem and are prepared to listen while we work out what it is, have our feeling witnessed and move on. If, for whatever reason, this becomes disturbed and its natural flow interrupted, sometimes by one or other of the parents, sometimes by some other event or even an unfortunate mismatch, then we get stuck until that need is met. Most people do the best they can and no counsellor *knows* exactly what it is that made their client get stuck. We are all so unique. One child's delight is another child's horror. The purpose of counselling is not to blame but to heal. The point about this is that *the trauma can be healed if the needs that were not met as children get met, instead, by the counsellor in the counselling setting.*

As adults we can take responsibility for meeting our own needs, but it's not easy to listen to yourself sympathetically if you really don't think you're OK. Your counsellor can provide those things that I have looked into in this chapter. He or she can give permission, be a benign mirror and a benevolent authority. He or she can really see you and really hear you if you're prepared to show yourself and in order to go with this suggestion we have to *trust in the process...* How I dislike this jargon. To someone in severe emotional pain it sounds so superficial. From the lips of any successful professional it makes me want to scream that they obviously haven't got a clue what I'm going through to talk in such clichés.

I do not mean to be glib. To trust in the process you have to go at your own pace, perhaps a bit at a time, perhaps round and round in circles to prove to yourself that this counsellor can be relied on *not* to abuse or ignore you. Now I'm going to use one cliché I *won't* reject: It'll take as long as it takes.

What Makes 'Help' work?

Lots of things, but some of them *have to be there* for counselling to work. For instance, if we don't have agreements about when and where to work then we don't have a place where we know we won't be interrupted. Therefore we will not feel safe enough (with very good reason) to experience what may be truly intolerable feelings. We need to know *when* we are going to see this person who knows our deepest darkest concerns, again. That's why counsellors can seem so rigid and, from the outside, quite prissy about appointments, and why such time and care is taken about making agreements. We also need to know *how* the counsellor will be, so that we can begin, with time, to trust them.

- We need them to be empathic, and be able to hear what we are saying or trying to say.
- We need them to accept us as human beings worthy of respect with valid points of view regarding how they see things.
- We need them to be able to be real people, with reactions and values of their own, as long as they don't foist them upon us unnecessarily.

Parents are God-like in the power and (often unintended) influence they have over their children, so it's unfortunate that people are so bad at providing the unconditional love that we all need. Children are so small, so unformed: parents are so big, so loud, so all knowing. If they don't see who we are we're more likely to decide we don't exist than to decide that it's their fault. I remember distinctly catching a train with my mother at the age of about nine and thinking that she must be omnipotent because she knew which platform to change at. No wonder this level of adoration can come crashing down during the teenage years!

How do we moderate this state of affairs? We have already seen that we don't actually need to be God on earth, we just need to be 'good enough' parents. As we move children towards the realisation that their parents are fallible we teach them humanity. It is often the first of many illusions to be discarded. There will be so many more. To deny anyone the truth is cruel, but to deny our mistakes and never apologise to our children is worse.

The same is true of counsellors. They too have to be 'good enough', so that imperfection and disappointment can become a tolerable and realistic part of life without it being the end of the world. They may have guessed how we are feeling or where we are at and got it wrong, but at least there is an 'us' left to evaluate that. When we too, sometimes get it wrong again, we are not destroyed, just imperfect. A true reflection and one which is ultimately acceptable. The good enough counsellor strengthens and nourishes us enough to hold up a mirror to our highest as well as our most fallible selves, and shows by example how to accept that with grace, in our completeness. Which brings us back to Maria, whose stored up favourite things seem to make her big enough to weather any disastrous storm. How I always wished she was my mother!

Exercise

For most of the exercises in this book, take five minutes to prepare. Make sure you will not be interrupted by the phone or the door. Stick a notice up and/or tell others in the house not to disturb you. If this is not possible, wait and choose a time when it is. Gather anything you may need around you. Tissues, a comfortable pillow, pen and paper. Allow yourself to create a gentle environment, perhaps with candles and aromatherapy oils. If you wish, record this exercise onto tape, reading at your own pace. If you are over 16 and you want to do this but it is never possible, make an appointment with a counsellor immediately as you are somehow denying yourself or being denied your rights!

Grounding

1. Place your pillow or cushion on the floor and sit on it in a way that feels comfortable. This may be with your back against something, with your legs drawn in or outstretched.
2. Notice what feels comfortable. Allow time to make a decision about how to sit. Weigh up the pros and cons of the different positions, bearing in mind that you're going to stay awake and not drift off.
3. When you are settled, close your eyes and feel the cushion beneath your bottom and legs and beneath that the floor. Let the lower half of your body feel heavy as lead, anchored to the earth.
4. Imagine invisible energy streaming away, through the floors, to the foundations of the building you are in. Imagine it going on through the topsoil, the subsoil and eventually the rock of the earth. At the centre of the earth is a burning fire. Stay by it for as long as is comfortable. Imagine the energy of the burning fire coming back, through the rock and the soil, up through the building to your feet, your legs and, eventually, your solar plexus, where your tummy button is.
5. How do you feel? Is there a colour or a phrase that comes to mind? Write it down. There is no right or wrong answer to this. Perhaps you found the exercise impossible or you feel very connected. Whatever you feel, try to get as close as you can to describing it, even if that would make no sense to an onlooker.
6. Stay with the feeling, even if it is uncomfortable, until you have an accurate enough representation of it and draw or write this in the middle of your piece of paper.
7. This feeling is likely to have lots of other feelings, beliefs, judgements or thoughts that go along with it or are tagging along behind it. Write or draw them as sparks flying off the words or picture already written down.
8. What you have there is a tiny and unique map of what was uppermost

in yourself for that brief moment in time. It will always vary, as you yourself vary.

9. To finish off, thank the earth for its energy, feel yourself sitting solidly and appreciate the way you have just given yourself the time and space just to be.

10. Keep your map and acknowledge it as part of you. As you use this exercise more and become familiar with it, use the question what do I think about ….(current issue for you). Travel again to that fire and see what it has to tell you. Do not judge your response. Merely take note of it and keep it small and manageable.

Chapter Three:

What We Don't Need

'And she could see no reasons
'Cos there are no reasons
What reason do you need to die?'
The Boomtown Rats: I Don't Like Mondays

In 1985 Bob Geldof opened Live Aid to a global audience of 5 million with this impassioned howl of anguish. He paused, holding the moment so that we could all connect with the incomprehensible fact that thousands of children across Ethiopia and Sudan were dying of starvation at that moment. *I Don't Like Mondays* is a song about a schoolgirl from San Diego who, one Monday morning, leant out of her bedroom window and shot 8 people. When asked why she responded, 'I don't like Mondays.' Incomprehensible? To the adolescent mind so is the fact that governments cannot save children from dying of starvation in a world full of food, a fact that may seem as cruel and violent to the schoolgirl as her actions do to us.

Counterculture tends to side with the underdog. By definition, it knows what it is to be the underdog. That's why the Boomtown Rats name is self-explanatory. It says: *'We are punks. We can't sing and we can't really play. We're on the bandwagon and into the money.'* To be an authentic punk rocker you couldn't be a hypocrite. You had to see and say how it was. In this, punk could not avoid being political, though it was not dogmatic. Never dogmatic. Punk was about trying to get to something real and if the only thing you saw going on was exploitation, rage, greed and violence then it was better to expose, parody or be it than pretend it wasn't at the core of society.

Inevitably, punk was youth – no strings, no experience, no future. Punk was a sort of performance art making the truth about reality as the adolescent saw it bigger than reality itself. This is what art does. For example, a good film about taking heroin gets closer to exploring that experience than simply filming someone taking heroin. A good film *about* taking heroin expands awareness, whereas the experience of actually taking it makes people understand *less*: it diminishes awareness.

Trainspotting demonstrates this admirably. It doesn't preach or analyse. It doesn't need to. Renton (played by Ewan McGregor) is an anti-hero who is anti-life because, in his *experience*, life means:

"Sitting on that couch watching mind-numbing,
spirit-crushing game shows stuffing
junk food into your mouth"
Trainspotting

In other words, life means consuming rubbish, swallowing junk. Anything but a hypocrite, in the true spirit of the counterculture, Renton is committed not to saying how it is, but to actually doing it how it is:

"I chose not to choose life. I chose something else.
And the reasons – there are no reasons.
Who needs reasons when you've got heroin?"
Trainspotting

Perhaps we understand that this is the same rage that Geldof was expressing at Live Aid, a directed and matured outrage that led to the feeding of half of Ethiopia. It is motivated by sensing that there is nothing to understand except profit and greed. Renton knows his reasoning only too well, much as he might want to numb it. He kicks his habit, hoping for better things but only gets a clearer view of how bad life is:

"It's shite being Scottish.. We're the lowest of the low.
The scum of the fuckin' earth. The most wretched, miserable,
servile, pathetic trash... Some people hate the English.
I don't. They're just wankers. We, on the other hand
are colonised by wankers. We can't even find a decent
culture to be colonised by. We're ruled by effete assholes.
It's a shite state of affairs to be in Tommy and all the fresh
air in the world won't make any fuckin' difference."
Trainspotting

Immediately following this outburst of painful clarity, Renton goes back on heroin with a vengeance. As he narrates over endless images of burgling and running, this is hardly a soft option. As he says, financing a commitment to heroin addiction is a full time business.

Responses to this film were deeply divided between distressed concern that the use of heroin was being glamorised and deep joy at the way the film exposed taboos previously underplayed:

> *"People think it's all about misery – they forget the pleasure of it. Otherwise we wouldn't do it. We're not stupid.".*
> Trainspotting

Renton, Spud and Sick Boy are at no time more fully present than when they're running from the scene of the crime. The *Underworld* soundtrack pumps the word *Lager Lager* through our veins like the heroin they're about to shoot into theirs, proving, without a doubt, that we, as well as they, are alive: fully, painfully, excitingly alive.

Those of us close to the adolescent mind that feel this buzz are fed and strengthened, recognising something we do understand at last out of that huge confusion that characterises the teenage mindset. Those of us that are repulsed feel alienated and weakened. The adult mindset necessarily doesn't understand and frequently reacts by rejecting behaviour outright as if the nasty bits of a counterculture are somehow incidental. In fact, the foul language and the filth in *Trainspotting* is as important as the safety pin through a punks nose because the counter in counterculture is as essential as the *turning* in turning the tables. If the mainstream isn't made to feel as alienated, sick, angry and disgusted as the minority do most of the time then the counterculture has lost its balancing function and society has lost an empathic dimension that is crucial to its motion.

Deprivation

So is Renton mad or bad? This is one of our favourite questions as we search desperately to distance ourselves from such antisocial behaviour.

In the musical *West Side Story* the Sharks (a street gang) sing a song to the policeman who is always on their case. As they sing they act out every expert's diagnosis of the problem. The judge decrees they're crazy; the psychiatrist says they drink; the social worker explains their behaviour as 'lazy' and the jailer says they stink. The kids know better. They know that the problem is, they're growing and they don't like what they see. The problem now is they're grown. Never mind social do-gooders who think they can solve their problems by calling them alcoholics or lunatics. In a line that comes straight from their pleading hearts they try once and for all to tell Officer Krupke:

"We've got troubles of our own."
Stephen Sondheim: *West Side Story*

Renton certainly has his own troubles. 40 years on from fashionable left wing ideas about not crushing freedom of expression, our 'deprived' kids have more material wealth, greater right to complain and more charters, acts and laws on their side than they can reasonably spend, act on or understand. With X boxes and Playstations accessible to all, designer trainers on their feet and mobile phones coming out of their ears, it's clearly time to separate the reality of the 'deprived child' from the myth. Deprived of what exactly?

Well funnily enough, perhaps they are deprived of reason. Perhaps it is no coincidence that Renton's mother is a valium addict who is simply not available. Physically she's there but mentally she has no sense at all of who he is, how he feels, or what he thinks so he simply cannot be argued with. He is therefore just not given the means to reflect that his view is perhaps selective or not the whole picture.* Renton's mother feels sorry for him and my guess is that he despises her for it, hating her for not providing an anchor or a firm, more considered, perspective to add to, and therefore modify, his own.

It's an insane world, as Renton illustrates with such eloquence throughout the film. Totally insane? For the adolescent, everything is total. It is the job of the parent/adult to teach that things are not black and white but grey and complex, but do we do this? Like Maria in *The Sound of Music*, Renton is of course, a symbol. There are very few

* Erikson (appendix) shows us some of the challenges we need to meet in order to grow into adults.

heroin addicts who are really like him any more than there are many all-singing, all-dancing, 'made-for-motherhood nuns.' It is the madness of the world around him that Renton sees and responds to and this vision, of the world as mad, is what we have to grasp if we are to understand the type of behaviour Renton and his mates display.

In the old days, over-indulgence would have been blamed for a lack of inner discipline which in turn would have been blamed for children's bad behaviour. Now, as consumer goods threaten to drown us in packaging alone and with the myth of over-indulgence the eternal off-the-shelf explanation for bad behaviour, it is harder than ever to consider that deprivation could be the problem. We swing wildly, providing kids with material things and then, when they misbehave, we call them 'spoilt'. Could it be that they are deprived of reason because we behave in such a silly way, providing them with food, sweets, TV and traffic overkill until their natural reason and needs are swamped? On top of that, we tell them they are what we want to believe they are and if we don't like their response there is every chance we will tell them 'you don't really feel that'! No wonder the kids don't know if they're coming or going.

Imagine a betrayal by God on such a scale that the most beautiful things on earth were going to be taken away and replaced with poisonous rubbish. Consider Loch Lomond emptied and quarried, the surrounding hills covered in tarmac, the sky above full of ash. It would be almost impossible to accept because it would be against all reason. Yet from the child's point of view, God is surely the grown up, the source of all reason, so if God did such a thing, what would be the point of anything? If the sun rose one day and it was black, we might all be tempted to slit our throats in despair. Within that context, a context of no hope, that would be a reasonable thing to do.

To understand the behaviour of disturbed youth in our culture we have to consider that by not allowing them to be heard properly, we leave them very limited options for the ways they can behave, because their behaviour is an expression of what they need to communicate. In fact, in the face of the black sun example (above) it really is admirable and astonishing that they do not slit their throats and what is a government that allows children to starve if it is not the equivalent of a black sun? Deprived teenagers find endlessly creative ways not only to express their experience but also to live with it. In *Trainspotting,* it is uncanny how the dialogue around withdrawing from heroin would serve just as well if the missing substance was love. Consider feelings of bereavement:

"The sickness is on its way. Sweats, chills, nausea,
pain and craving. Need like nothing else
I've ever known will soon take over me."
Trainspotting

Renton is empty inside because he has never really become anyone. Still struggling in his late teens to be properly responded to, he is so desperate for confirmation of what he sees that he is even hoping that someone will lock him up and finally treat him appropriately. That is, in response to who he really is: by this time, a thieving heroin addict:

"...propelling ourselves with longing towards
the day that it would all go wrong because no
matter how much you stash or how much you steal,
you never have enough. No matter how often you go out
and rob and fuck people over you always need
to get up and do it all over again."
Trainspotting

And we say there is no reason behind his behaviour? If *'Can you see me?'* is a central question which the development of the child is dependent upon getting an answer to, perhaps there is meaning, even though the reasoning seems warped. Clearly, it has taken Renton some time to become this desperate but I suggest that the question remains:

Can you see me now?

Just as having our needs met gives us a basis for believing positive things about ourselves, so not having them met gives us a basis for believing negative things. Our reasoning unfolds from these beliefs, and astonishingly, like a flower, nearly always still attempts to go towards the light despite having to warp and bend around insurmountable objects.

Must Try Harder

In the previous chapter we looked at that perennial question 'Am I alright?' and the way that people look around for confirmation that they are alright. When things have been really bad - when the 'good enough mother'* has not been available – what we look for is confirmation of the belief 'I am a really bad person'. In the last chapter we recognised the relative health within the freedom to say 'I hate my mother'. The kids we're talking about here don't say 'I hate my mother', they say 'I hate myself' and 'I am evil.' In psychology we say the message has become internalised**, though it is not particularly likely to be conscious. The kids then swing wildly between hoping this is not the case and despairing that it's really true. Ultimately, to believe that you really are bad is much more tolerable than to believe that the pain and anguish inflicted on you was really done by the person who is supposed to care for you most. In a more moderate setting this is what occurs when kids fantasise that they are really princesses, stolen at birth! Then it all makes sense! Gradually, as we grow and begin to tolerate imperfection, the fantasy fades and we accept the reality: Life is difficult.

Here though, we are not talking about what is moderate, but the effects of consistent abuse or neglect. In these cases, the mirror referred to in the previous chapter teaches worthlessness. If the child who looks to the parent or carer for positive feedback gets spat on, used or ignored before they have really developed a sense of their own value, they will develop an unconscious belief that they must be terrible to be treated in this manner. Inevitably they search both for the approval they need by *trying harder* and the unconditional acceptance they need by acting out how they really feel. The desire to be really seen remains, which is why these kids oscillate between being 'good' and being 'bad'. Until someone properly understands their behaviour and the kids know it has been understood, they will just keep on doing it.

As life began for these kids, so it continues. They hope to be seen and accepted, but they know, that is they believe, (just as they have always feared and been right before) that they will be misunderstood. So they jump from wanting to be seen to not wanting to be seen, believing that they are entirely unacceptable but always hoping that someone will love

* Winnicott (see appendix)

** Instead of the message coming from outside, they start to tell *themselves* they are worthless.

them. Perhaps what is particularly difficult to grasp is the connection between deprivation and the behaviour of the deprived. If the Rentons of this world were a bit more grateful for some of the support structures that the taxpayer provided, that would fit with the adult sense of justice, rather like the old notion of the deserving poor. However, if God were to put out the sun and then replace it with a 60 watt light bulb, which is what life looks like if you really are deprived in the sense that we are discussing here, you would more likely be angry than grateful, and as that famous adolescent *John Lydon* of The *Sex Pistols* said later in his career:

'Anger is an energy.'

It would be better than despair.

* * *

The Rentons of this world are often those who have been both brought up in relative poverty as well as suffering from emotional neglect and abuse. Many of these kids end up in care and because of a lack of specialist help, a huge percentage at some point in their lives become homeless, criminal or ill so they come up against institutions and the authority of those institutions again and again. The impact of this perceived and very real power is extremely important. It needs to be a power for good – a part of the solution but for the following reasons it frequently reinforces the problem.

Confused Agendas

It's no coincidence that two great bands of the 1990s had the names *Arrested Development* and *Rage Against the Machine*. During the nineties it often seemed that those state supported machine-like agencies who appeared to have a mandate to reverse the experience of disturbed people often succeeded in driving the problems deeper. There is a parallel here with the principles of homeopathy, which hold that if you treat only the symptoms of physical illness, you drive the illness deeper. In the care system, what is so easily driven deeper is the conviction in the clients that just like everybody else, Social Services don't care, can't be trusted and aren't really listening.

This is down to a basic conflict between what Social Services appears to be set up to do and the complexity of its agenda. Given the rhetoric and the policies put forward it is easy to assume that the psychological welfare of children in care would be the priority. In reality the long term welfare of the child frequently comes second to:

- Considerations of the budget for one department rather than another
- The wishes of the parent, however they may have behaved in the past
- The needs of reorganisation, and social services is always reorganising

The outcome is that children who are disturbed and struggling often have to accommodate the needs of the social work organisations whose main aim is to meet the needs of the children!

Broken Agreements

The systems of the helping services suffer from the same complexities as many large organisations. Within many voluntary projects and social and psychiatric services, appointments are frequently not kept, rooms are double booked and, due to disorganisation, lack of resources and lack of training, agreements are often broken. Social Workers, often worn down over the years have become used to the possibility that therapeutic help may get interrupted or appointments cancelled, missed or even cut off halfway through a series of sessions. Unqualified helpers are simply unaware that cancelling a session has the effect of smashing people's trust again. While 'listening support' is a skill that can be learned on a weekend course or over 3 terms one afternoon a week, an understanding of the realities inside these clients' heads will take a good deal more study and experience, and this type of training is no longer even on the social work training syllabus. While agencies claim that their intention is to empower children, it is often the case that they are completely unaware of how they and the system endlessly undermine the very thing they are trying to achieve. Because the 'helper' has the power to do this and the 'helped' does not, the realities they occupy are essentially different and all offers of 'empowerment' to the 'helped' remain empty words. It is remarkably easy to do more harm than good, reinforcing the clients belief that you can't trust anyone. They have found the evidence that their assumptions are true (I'm a bad person and not worth helping) and we, who are supposed to be helping them have not prevented this from happening. Then their anger at being let down makes them look crazy and their guilt makes them try harder until they really blow up and do some serious damage, generally to themselves. As far as the 'helped' can see they were right:

"Everyone just wants to get rid of me because nobody cares."

In the same way as the rich get richer, the abused get more abused, picking up blame like fluff sticks to nylon. One of the confusions here lies in the use of the term 'counselling'. What is called 'counselling' within the limits of these agencies is really 'listening support.' Counselling is an activity which at its core addresses the attachment issues of contact, betrayal of trust, clear boundaries, intimacy and abandonment although most of the time the conversation going on in the therapy room is not directly or consciously about these things. To not understand this is often to re-abuse those clients who most need a secure place to work with their feelings.

Unpredictability

Young people who are struggling to make sense of a world that they experience as chaotic, untrustworthy and self–interested need to have an experience of the world as better than that so that they can begin to trust. They need a team of carers who are aware of what they do, who will challenge them when they break the rules and will act to hold them tightly when they seem to be straying into danger. The kids themselves devise logical rules for the carers. If they have been told they will be 'looked after' then, often against their instincts they attempt to trust by expecting caring responses and appropriate reactions. Then, as previously abused or neglected people tend to, *they watch for evidence that the carers won't bother*. It only takes a little forgetfulness or a lack of awareness of this process on the part of the carer for the kid's idea of the world's total self-interest to be confirmed. When agreements get broken, kids feel deeply betrayed and all the rules go right out the window with understandable contempt.

Knowing what is really going on takes considerable expertise so good care for disturbed clients is expensive. One young girl I worked with had her 'specialist'placement ended because her behaviour was better and she was 'manageable'. Having got over many of her struggles she was preparing to study and looking forward to progressing in life. When she was told her placement was ending she naturally thought it was a punishment for being bad! As this young girl's counsellor I felt that for the sake of her sanity, she deserved the truth. That actually she had been too good! And we question their ability to reason? Why, in her situation, would she feel that she owed anyone anything?

"This was a boomtown where any fool could make cash
from chaos and plenty did. I quite enjoyed the sound of it all:
profit, loss, margins, takeovers, lending, letting, subletting,
subdividing, cheating, scamming, fragmenting, breaking away.
There was no such thing as society and even if there was
I most certainly had nothing to do with it."
Trainspotting

Bright kids see the injustice of the world around them and respond in kind to what they see. Until we start both to understand that and to *prove* that we understand it, how and why on earth should we be able to facilitate respect for and trust in authority?

* * *

Although those who have been battered and abused may have a shaky sense of how to behave, they certainly have a powerful sense of justice. For those people who are not consciously aware of their rights this can lead to all kinds of distorted ways of getting them. The only thing that's changed is that:

'From now on we play by my rules.'

From the perspective of the kids, the actions of many 'people with problems' can be seen to be remarkably sane. I knew one young man who, by the time he was 16, already had a criminal record, had become a heroin addict and had burgled houses up and down the road he lived on as well as kicking down the doors of those he shared a house with. Nothing was done about this. Next he came to the office of the organisation that housed him where he took the opportunity to steal car keys from the bag of a worker and drive their car out of the car park. This was witnessed by those who worked there. What happened? A policeman took a statement and went round to the young man's house but he was out. The next day I telephoned the police to let them know that the young man was in. Their response? That the officer who had taken the statement was away for a week and it would be dealt with then. I think we would have to admit that to this boy's deafening statement:

"Is anybody listening? Is anybody out there? Does anybody care enough to stop me?"

The answer was definitely 'no'. I would go as far as to say that sometimes the messenger may be extremely crude, even violent, but within the context of some of the kids in the care system there is a terrifying sanity to their behaviour.

This level of youthful vitality is the life force that inspires us. It should never be disrespected. Challenged, yes. Sometimes locked up as long as therapeutic concern is being prioritised, but it should never be seen as not being of value. When Spud takes some amphetamine and heads for his job interview, his delight, both in the language and the insights stimulated by the drug, says it all. When asked what attracts him to the leisure industry, he responds:

"My pleasure in other people's leisure...I'm a bit of a perfectionist. It has to be the best or there's nothing at all. If things get a bit dodgy I just can't be bothered."
<u>*Trainspotting*</u>

Tolerating frustration is one of the central difficulties for these kids. As seen in the previous chapter, it is a learned skill. It takes adults to show the way (and how good are we at that?) Listening is not enough. We have to hear, to accept and to respond. We have to see what their behaviour is telling us. Then we have to provide the consistency that we know they need in order to help them through. A recent Channel Four documentary called *Brat Camp* was a perfect example of the way in which, given support in their struggle, teenagers can overcome their own massive blocks to growing up.

No Breakdown Cover

It's not all bad news. Mental health services can be massively helpful in that they can provide a temporary refuge and asylum from the demands of an unmanageable world. Just the opportunity to be away from pressures of family, work, and bills can be enough to precipitate a breakdown.

To precipitate a breakdown?

We tend to assume that we go into hospital to take pills and get 'fixed' but what if instead of a breakdown being a fault it is the thing that needs to happen to enable those who are suffering to recover? What if it is a way of bringing us back into balance? We see how bad it can get and then we start again. If we had the money and the social skills, we could go on holiday to de-stress and get some perspective. Increasingly, people can now attend holistic health farms with talking and alternative therapies, taking responsibility for their own mental health in a way that would have been unthinkable three decades ago. They let it out, crack up, eat well, get put back together again, reflect and get stronger. On the whole, these people are relatively wealthy, articulate and successful people. They have the opportunity to approach balance in their lives in a highly sophisticated way. The people in *Trainspotting* differ only in poverty of experience, wealth and aspiration. They approach the balance in their lives unconsciously, which is as well since if they walked into a psychiatric hospital and asked for asylum they would be denied it on the grounds that they were fine. This, of course is the central tenet of *Joseph Heller's* famous novel *Catch 22*. The main character *Yossarian* wishes to be grounded to avoid flying on a dangerous bombing mission in World War Two so he pleads insanity. And here's the catch - how could anyone who was insane make such a rational and life preserving decision? He must be sane, therefore he has to go and bomb people.

Just as health farms for the well-off help people to consciously get what they need, so eating disorders in young people can be about unconsciously getting what's needed. Absolute control over what goes into your body as well as when it comes out, and how, is a response to the inner feeling of being in chaos. How can everything be that chaotic when you exercise that level of control? Certainly there's a much better way of living but let's not dismiss *their* attempts. The anorexic's body becomes an illustration of the deep ambivalence between wanting to be looked after and wanting to be left alone. It shows up the chaos and tries to overcome it all at once. From their perspective it's an entirely sane response, and the best they can do. It's as clear as the nose on your face if you've got the coded language, but you've got to really understand what it is to see it through their eyes.

Self Analysis

The last thing that we need in these serious and potentially life threatening situations is analysis. Perhaps this is why the older generation is sometimes so much against counselling. They think it involves 'navel gazing': an endless working out of increasingly outside chance reasons as to why we behave as we do. In fact, this is not counselling but analysing oneself. There is a huge difference. Analysing oneself almost always begins with asking ourselves 'Why'? which goes straight to the head and our limited ideas, rather than to the heart and to the feeling. It is a popular misconception that counselling will lead to being scrutinised, judged, diagnosed deficient and dismissed. Often clients are terrified of what I am going to tell them, but come because they are desperate. Equally many people don't come because they don't see how me analysing them can help since they know so much about themselves already. The point is that their view is restricted to what they already believe about themselves, and 'why' offers no real tool for getting beyond that. In counselling, the counsellor rarely asks why. Most of us already spend huge chunks of our lives justifying and defending our behaviour. Counselling is about trying to understand and places the emphasis more on what's positive about behaviour, or what it might be trying to express.

Generally, a client's deepest fear is having their deepest fear confirmed so apart from the 'why'question being rather accusatory and making us feel cross-examined, the very fact that somebody with power is focusing on a particular topic can instantly scream *'problem'!* This confirms our fear that something is wrong with us and that if we can only find the right answers to these questions, the problem will be solved and we'll feel better. Before long 'analysis' has taken a path not to the beach for a relaxing swim, but to the smelly and toxic rubbish dump because before we have even begun to explore what it is to be experiencing this difficulty, we are rolling about in the unacknowledged message: *Yes, you are a very disturbed/bad/wrong person* and its even more alarming solution *but the expert knows more than I do and can help me change.*

Bearing in mind the self image and the beliefs we, at least partially hold about ourselves, are these assumptions healthy, or do they in fact, dovetail with our already low self-esteem to grind us further and further down? The problem with assumptions about one's self or about how things are is that you can only step outside them when you see that they are not 'fixed reality', just thoughts. Until we become aware of that, we look at the world and ourselves through the eyes we've been given. Once we are aware of this, the statement 'I feel I should be able to deal with this myself' is often not a statement of brave intention but is a statement about how stuck we are in our belief system. What then occurs, when we don't deal with it, is confirmation of our *belief* that we are a weak and worthless failure, just like we knew all along!

Look at it this way. If someone gave you a field, a fork and a load of weedkiller and said 'create a garden', you'd be a bit limited with what you could do. Some sun, rain and seed would be necessary. If someone gave you some crayons, paper and a pen and a load of self-loathing, you'd be hard put to come up with anything to lighten the heart. You would need some love and another perspective on yourself – and I don't mean just being told that you're alright. I mean being treated as alright. At the end of this chapter is an exercise to start us on the path to seeing how we are thinking and the messages we give ourselves. Perhaps when we can see our thoughts clearly we can take steps to stop them from dominating our lives. This is of course, a lifetimes work and this process is one we will return to again and again throughout the course of this book.

Self Help as Self-Annihilation

So – how much time do you spend cramming what you learned from self help books down your partner's throat because you've now learned that it's like this or like that? Have you told them they are 'projecting' their inner problems on you? Do they have a damaged 'anima' or 'animus' and that's why they behave in that way? Glamorous isn't it? Yet ultimately analysis is like a Fisher Price toy to a baby: fascinating and dazzling. Rather like a charismatic man, it may not go the distance.

When we get involved in doing this stuff we're often more interested in scoring points and shoring up our self image; trying to become stronger instead of feeling weak and helpless. It means that when we look in the mirror that is our partner and don't like what we see we try to shift the blame onto them. They are not 'evolved' enough, they don't 'try' enough. Then we shift the blame onto ourselves and find ourselves fighting against our own beliefs and the imagined beliefs of our partners and it feels like life or death. Then we try harder and harder to be 'good', to be *more* evolved, to try *even* harder, driving the assumption that we are 'bad' deeper. Perhaps it is true that we create our partner's opinions in our own self-image: that is, we imagine their criticisms of our behaviour and try to avoid what is actually not there at all. The problem is that if you already live in an inner landscape of guilt and self-hatred, analysis is just going to draw you further in. Discussions filled with new age platitudes such as 'well – you create your own reality' will do the same. What is needed is to feel and get real, and to find some compassion.

In *Trainspotting*, Renton is a broken idealist who becomes a realist when the mould is smashed by some new and extreme experiences.

Renton has known all along that the junk food, the adverts and the game shows are worthless as a substitute for the love he lacks, but it is very difficult for him to see that he is more than he has been shown. It is only when he faces the death of friends, the persistent violence and abuse by Begby, the opportunities of London and some serious money that his perspective changes. This enables him to take steps towards a new type of life.

The change in Renton's circumstances, these new and extreme experiences, the awareness perhaps that this is possible – *that life is not fixed but fluid and changing*, enables him to understand that he is free: in particular, free to choose what things mean to him, that the adolescent mindset is only a part of the picture. He is free to define how and when to use things, instead of them defining him and his life. *Things no longer mean what they meant before.* His decision to take the money liberates him from his peers and from his peers' restrictive views but that doesn't make him entirely immoral, just appropriate – hence he leaves a large wad of cash for Spud.

For Renton, deprived but bright as a button, things have never meant what they meant to others, and in his non-conformism, he challenged what was usually unconsciously accepted. This led him to the possibility of growth and of living another way. When he says at the end:

> *"The truth is that I'm a bad person but that's going to change. I'm going to change. This is the last of that sort of thing."*
> Trainspotting

He speaks with the only words available. He was 'bad' and now he is 'good'. What he means is that *all* of it is rubbish. What he's looking *with* as well as what he's looking at: both *his own loathing and the worldview that developed with it*, and he knows it. Once we know something, we have choice. Renton becomes liberated from his fixed definitions, including his own view of commodity as a love substitute, and finds that life offers all sorts of (fun) aids to living that maybe can just be what they are: whitegoods!..

"I'm cleaning up and moving on, going straight and choosing life…The job, the family, the fucking big television, the washing machine, the compact disc and electrical tin opener, good health, low cholesterol, dental insurance, mortgage, starter homes, leisurewear, luggage, three piece suite. DIY, game shows, junk food, children, walks in the park, nine to five, good at golf, washing the car, choice of sweaters, family Christmas, index pension, tax exemption, clearing the gutters, getting by, looking ahead, the day you die."
<u>Trainspotting</u>

Maybe, as the best we can do, it's not so bad. We don't have to get trapped or deceived by it. So – happily, Renton has finally grown up and seen that the world is not black or white or even chequered. He has accepted something of its and his own imperfection and become much freer because of it.

Exercise

Permission: 1

Initially this exercise may be quite demanding and it is also potentially harmful. For this reason parts four, five and six of this exercise are more important to do than the first three! Learn it only when you are comfortable and have some time. Initially do it only in the 'safe space' that was described as essential for the first exercise. You will need a box to keep your work in (a shoebox is an ideal size), a pen and some long strips of paper, half the width of A4.

1. Think of a recently stressful time. Keep it small. A stress about being late or forgetting something at the shops will do. Notice what you were thinking and identify it. If it escapes you, try and remember what you were thinking about a few moments ago.
2. Was it a command to yourself (an 'I must')? Was it a criticism of yourself (an 'I'm hopeless')? Was it a criticism of others? Was it an appreciation of yourself or others?

3. Distil it into a phrase. This phrase might begin "I must...." "I should....." Practice this part of the exercise until you can do it easily. When you feel that this distillation of the message really conveys the meaning of what you are telling yourself, write it down at the top of one of the pieces of paper. Do this with as many pieces of paper as feel right.

4. Take the paper you have written this on and prepare to fold it so that the message cannot be seen. Imagine any harshness connected with this message as the colour red. As you fold the paper, imagine all the redness being drawn and packed neatly inside.

5. As soon as you have a chance, put the message into your box and then wash your hands. Imagine yourself washing away any of the red harshness that lingers.

6. Only take the messages out of the box when preparing to do further exercises in this book and don't leave it where anyone might intrude upon it.

Later on, when you become familiar with the process, you can do this exercise anywhere. In the loo, at work, on the bus or in the car. The more you do it the better you'll get at it. Choose a word from words you use everyday such as 'bus', 'key' or 'bread' to remind you to 'tune in' to what and how you are thinking. That way you can stop and 'catch' your assumptions to become aware of how and what you habitually think.

Part Two
The Preparation: Heroism In Practice

Chapter Four

Ideas To Discard

Girl giving out leaflets: "Have you accepted
Jesus as your personal saviour?"
Buffy: "You know – I meant to and
then I just got really busy"
Buffy The Vampire Slayer

With one quick fearless stake-in-the-heart, our heroine shows why she's so great. She doesn't just slay vampires, she slays all the myths that prevent us from being who we are. Buffy constantly uses flippancy to treat dogma with the superficiality it deserves, undermining the 'whole big sucky thing' by refusing to buy into the fear it generates. The real drama of being a teenager is enough and that's what Buffy is really all about.

' One girl in all the world... with the strength and skill
to hunt the vampires, to stop the spread of their evil... '
Buffy The Vampire Slayer

Like all true heroines, she's reluctant. Unable to avoid her destiny or her moral obligations, to be Buffy is to be as free as it gets. She doesn't stand in disbelief that vampires exist, she comments on Giles weird obsession while it sinks in:

Giles (the 'watcher' who is the slayer's mentor):
"Werewolves, incubi, succubi: everything you've ever
dreaded was under your bed but told yourself it
couldn't be by the light of day. They're all real."
Buffy: "What, you like, sent away for the ten life series?"
Buffy The Vampire Slayer

And then, like Harry Potter or Frodo Baggins she gets on with it.

Buffy is all teenagers who search for meaning in a landscape of unknown terrors and inexperience. She is heroic because she accepts herself and takes on the full implications of evil in a complex world without hiding behind convenient lies or illusions. She searches, she struggles, she wins – and sometimes she loses. Any teenager or ex-teenager who has ever had their heart broken will know better than to disrespect that truth.

Lies Are OK

The main lie that Buffy disposes of is the one we all collude with as teenagers. If it stays unchecked, it becomes so convincing that we think it really *is* us. It says:

- I know fully what I'm doing, why and what will happen.

Because this always seems true of everybody else, we hide behind the pretence that this is true for us too. We pretend that things don't hurt, that we are confident, that we are 'grown up enough' to have sex, take drugs, smoke and drink. Buffy is different. Her responses to things are as genuine as they can get. She has even created a way of speaking that teenagers can use, giving them permission to be vulnerable, be more in touch with their feelings and speak from them.

"Nice to meet you – I think."

When she feels ready to have sex, she doesn't reject her own gut feelings, or her doubts, but uses them as part of the weighing up of options.

She knows that desire doesn't always know what is right:

"To act on want can be wrong, but not to act on want...
What if I never feel this way again?"
Buffy The Vampire Slayer

And Buffy's best friend Willow, instead of saying, 'just do it' or 'way to go' when Buffy has more or less decided that she will sleep with Angel (a vampire with a soul), just says, in the most sweet and quiet way, almost to herself:

"Wow."

For these girls, at 17, sex is awesome, and it is a big deal. Men like Giles (who is over 30) are not considered sexual in any way. Giles is paternal and protective as he should be. Outside the world of the teenager and their battles, everything is stable, normal and secure in that the struggles and hormonal changes are simply seen for what they are. Buffy's mother, in recognising this, provides the reassurance Buffy needs to wear her doubts on her sleeve; to be as authentically herself as anyone can be. Her mum says that that however hard she thinks about it or tries, she can't really understand her daughter:

> *Buffy: "I'm inscrutable huh?"*
> *Mum: "You're 16. I think there's a biological imperative whereby I can't understand you because I'm not 16."*
> *Buffy: "Do you ever wish you were 16 again?"*
> *Mum: "Oh that's a frightful notion. I wouldn't go through that again, not even if it helped me understand you."*
> *Buffy: "I love you mum."*
>
> *Buffy The Vampire Slayer*

In terms of developmental psychology, this is the perfect conversation. Mum relates to Buffy, respects her experience and keeps her distance all at once. Teenage girls badly need to say that the sky is green if it will give them a sense of themselves as *not-mother,* and that's precisely what they'll do if they can't get their difference and independence acknowledged. More alarming, they'll get pregnant, take drugs and run away to do it if they have to. It's no wonder then that Buffy is the opposite of a battered woman. In her way, she is as idealised a character as Maria Von Trapp. Both are heroines who can bend the rules, be flexible and respond appropriately to the unique set of circumstance presented to them, even when they make for a difficult and risky life. Both are women who live by their beliefs and their experience; both have *integrity,* both have had 'good enough mothering', which forms a core solid enough to fall back on. Buffy *does* sleep with Angel, but because of a curse (sounding alarmingly close to what happens in real life) he loses his soul.

"One moment of true happiness, ...one moment when the soul that we restored no longer plagues his thoughts and that soul is taken from him..."
Buffy The Vampire Slayer

As part of the storyline, what was an intimate openness is withdrawn, just like a man without a solid enough core freaks out when he gets really close, afraid that he will disappear:

"I gave her my heart but she wanted my soul"
Bob Dylan: Don't think twice, it's alright

When Buffy wakes up and finds Angel is not there the familiar male/ female pattern kicks in. She is not angry or annoyed, just worried about him, but when he returns, (without his soul) he makes light of it.

Buffy: "But you didn't say anything. You just left."
Angel: "Yeah- like I really wanted to stick around after that."
Buffy: "What?"
Angel: "You got a lot to learn about men kiddo, although I guess you proved that last night."
Buffy: "What are you saying?"
Angel: "Lets not make an issue out of it - OK? In fact, let's not talk about it at all. It happened."
Buffy: "I don't understand it. Was it me? Was I not good?"
Angel: "You were great, really. I thought you were a pro."
Buffy: "How can you say that?"
Angel: "Lighten up. It was a good time. Doesn't mean we have to make it a big deal."
Buffy: "It is a big deal..."
Angel: "I should've known you wouldn't be able to handle it."
Buffy The Vampire Slayer

A lesser woman would be entirely felled by such a response. That's his aim. Angel's response is typically macho, trying to ignore his own terrors by disregarding Buffy's experience. The assumptions behind what *he* is saying are:

- Sex is throwaway entertainment
- If there is conflict, the fault lies with the woman.
- Her experience and natural responses mean she is weak, inadequate and spoilt.

Adapting Gets You Loved

So where have we heard these myths before? 30 years of feminism have shown those who have been brought up to respect themselves that they are nonsense, but the more we crave love and have an empty hole rather than a whole sense of self, the more we compromise to try to get that love. Unfortunately, what we compromise and dilute is our belief, our knowledge and ourselves. If we're 'adapting' we are adapting to someone else's standard when we have to find our own. To try to cut out our responses in order to 'adapt' without understanding and sympathy for our yearning soul is to try and kill part of ourselves in an inversion of the true pain and uniqueness of being alive. As Angel said about Buffy:

"She made me feel like a human being.
That's not the sort of thing you forgive"
Buffy The Vampire Slayer

Often, in our attempts to be 'good' or to be what we think we *should* be we try to take a short cut, but this is like buying personal development off a supermarket shelf. Real therapy is about becoming liberated rather than fitting in for our own or other's comfort. It is about feeling our feelings and sometimes knowing wholly how it feels to not fit in and become more resilient for it.

Consider these beliefs that seem to be quite commonly accepted:

- We are responsible for our feelings
- We can control our feelings.

- Women are supposed to always care.
- Adults are never afraid.

While we still believed these were true, at least we knew what to do. OK, it meant lying to ourselves, suppressing our experience and forcing ourselves to behave unnaturally but at least we thought we knew who we were!

The problem is that these messages are not true which is one of the major reasons why trying to squeeze into them makes us feel so uneasy. The soul is not for sale or, if it is, what you get for it is not really worth having.

So how do I know they're not true? Because:

1. I can't control my feelings: I spent 40 years trying.
2. I don't always care.
3. I'm sometimes afraid.

I know because in spite of my culture telling me what I *should* do, my gathering and studying of that experience, my working with and embracing of that experience has confirmed how things are for me and, increasingly, how I *work.* I have checked and double-checked it, like a scientist. I know what is true *for me,* and what I've just noticed is I'm not spinning anymore, wondering what I should feel. I'm on solid ground, alone, just me and my experience. There are others out there, but they're not me. I don't know what they think and feel. I only know that their behaviour sometimes affects me and if I reflect on it, I know *how* that happens so I also know when I need to move away from it.

Clearly, Buffy has two choices when Angel starts to spin his cultural line. She can embrace the pain of being alone after such closeness or she can accept the lie that she is 'over-dramatic' or 'hysterical' or 'weak' and follow Angel around while he psychologically beats her up. Somewhere inside though she knows that the answer to the riddle 'When is Angel not an Angel' is 'when he's a shit'. Buffy may be vulnerable, but she's not naïve, and more to the point, her instincts are not injured. Unlike the battered women, whose need is so great that they fill their hungry-for-love hole with illusions and are often not whole enough to see what is happening in front of their own eyes, Buffy names abuse and is able to see that we each have good and evil, light and dark, inside us and we can't let the dark control us. Inside Buffy is access to a violence all her own. She needs it if she's to stay one step ahead of the game:

Angel to Spike: "Force won't get it done. You gotta work from the inside. To kill this girl you have to love her."
Buffy The Vampire Slayer

But Buffy is sure of what she loves and what she doesn't. She is so unable to pretend that what is hurting her so much could possibly be love or the man she fell in love with (even though it looks like him) that far from being disabled by need she is outraged by such abusive tactics (a paragon of health!) and uses that outrage to violently deflect them because she has an innate sense of what is healthy:

Angel: "You know, the worst part...pretending that I loved you. If I'd known how easily you'd give it up, I wouldn't have even bothered."
Buffy: "That doesn't work anymore. You're not Angel."
Buffy The Vampire Slayer

Angel's response is to tell her that *she* made him the man he is today!

Wow – feel the power and the guilt around that one. After all, even the Bible says it was all Eve's fault. This is the message that has been used for centuries to wrongfoot women and make them work even harder to be nice. The lie that:

- Women are responsible for what men feel and for what men do with those feelings.

Naturally it hasn't worked on our heroine. Angel's appalling behaviour (earlier he wrote '*was it good for you too?*' on the wall in a victim's blood) only helps her to see how he is manipulating her, doing everything in his power to make her doubt herself, feel guilty and feel responsible. Her entirely healthy, guilt-free response is to become righteously angry and therefore powerful, kick him in the balls, and when he jeers:

"You can't kill me can you?"

Instead of savaging herself for being soft, she is kind and gentle to herself, muttering beneath her breath:

"Give me time."

Assumptions Are Absolute

Perhaps Buffy is such a heroine because she can see things as they really are. She has no false assumptions to protect her from pain. She does not spend her time trying to twist her experience to fit with false beliefs in the way we have described. If she had, she might have ended up like a friend of mine. This woman was upset and angry because a lover had turned up two hours later than arranged, but because it was too difficult to be with these feelings she used the belief:

- I can choose what I feel

to turn all her anger and her blame on herself. She decided that if only she'd 'been happy' when he'd showed up, she'd have got what she wanted. This attempt to twist from what she *did* feel to what she believed she *should* feel meant that she analysed her behaviour using her beliefs and ignoring her direct experience. All this did in the long term was drive her need and longing for a partner deeper into the dark and move her further from herself. If Buffy didn't take her feelings and her hunches seriously, she'd be long dead. So what's different about her?

Maybe she doesn't need a dogma or someone else's idea of how it *should* be to cling to because she can hold onto herself without it. Maybe the ground of her being grows out of her felt sense and she trusts it. She is secure enough to let herself not know the answers, but know what she feels and she can see what is in front of her eyes. She communicates how things are for her without resorting to stereotypes or hysteria:

Jonathan: "I don't want attention."
Buffy: "No I think you're up in the clock tower with a high powered rifle because you want to blend in."

<u>Buffy The Vampire Slayer</u>

For Buffy, there is no solid ground other than this and everything is up for grabs, but because she knows what she wants, what she feels, has a *healthy* assumption about what is right (being free) and what is wrong (abuse) she can define and explain everything she needs by simply describing what is.

"Owen and I would like to be alone right now and for that to happen, you would have to go somewhere that's away."
Buffy The Vampire Slayer

She is astonishingly free of judgement, seeing problems as arising from things being out of balance rather than from herself not being good enough.

"A cranky slayer is a careless slayer."

Buffy is even able to avoid conflict within herself by recognising that two different things may be true at one time:

"Do you remember when you said I was like two different people? Well – one of them has to go but the other one is having a really, really good time."
Buffy The Vampire Slayer

We often generate our assumptions out of our early experiences and every thinking being has some. Sometimes they are healthy and sometimes they are bad for us, and we partially construct our reality from them. If we change them, our outlook and sense of freedom will also change. This is the basis of neuro-linguistic programming. Look around at any self help section in the bookshop and you will find reprogramming your brain to change your life marketed at you as if it was soap powder that will wash your grubby brain whiter than white. For me this mode of enthusiasm always generates instant suspicion which the NLP'ers would tell me they could 'root out' to make me more open, more loving and trusting. Perhaps it's the instant promises and the showmanship that makes it sound like a dogma to me, or perhaps it's the way in which emotion is not seen to play a major part, though the theory holds true. As a counsellor I am increasingly aware that my clients have very unhealthy, unconscious assumptions about themselves, and that there are some standard ones which are generated by abuse and neglect:

- I am worthless.
- I am a very bad person.
- No one can see or hear me.
- I don't exist and therefore can have no impact on others.
- If I try harder I will be loved in the end.
- I am responsible for everything that happens to me.
- If I think or say something, it will come true, so I mustn't.

As humans we are extraordinarily vulnerable, developing our beliefs to fit with our experience, making sense by gathering evidence to back up our beliefs. In one Buffy story where nightmares come true, Buffy's father arrives and blames her for her parents divorce because she was selfish and so much hard work. Since most children believe this is true, it hits her core belief like a repeat car crash. An unbroken leg will break and then mend, but a leg that has broken before will fracture down the fault line and be even weaker. Her powers fade almost instantly.

In another episode there is a girl who is ignored so much that she becomes invisible – an impressive metaphor of exactly how people feel when this happens.

These are learned assumptions, arising from an experience of being ignored, neglected or abandoned but Buffy is our heroine precisely because her assumptions tend to be heroic. Despite difficult feelings most of what she says arises from an allowing, non-judgmental attitude. When she tries on a short skirt and looks in the mirror she can say 'hello' to her reflection and call herself a slut as easily as she can wear a calf length dowdy dress and call herself a Jehovah's witness. The words become simple labels rather than fixed ideas that threaten swamp her with an identity she does not have.

She accepts that

- We expect stability but we don't always get it.

When Giles isn't really surprised that the vampires are back Buffy doesn't act cool, she allows herself both fear and annoyance.

"Well I was (surprised). It's my first day. I was afraid that I was going to be behind in all my classes, that I wouldn't make any friends, that I would have last month's hair."
Buffy The Vampire Slayer

Buffy is so aware of what it is to be normal that she is *extra*ordinary, but at least she writes large what health is. She seems to understand that assumptions are merely floating beliefs that help us make sense of the world. They are not who she is and they are not fixed truths. Buffy knows she's got them, she knows what they are, she knows that not everybody has the same ones and she knows they can change. She knows that depending on who or what you are, different things become taken for granted:

"It's safe to say that in his animal state (Xander has turned into a hyena) his idea of wooing doesn't involve a CD and a bottle of Chianti"
Buffy The Vampire Slayer

So how would we become heroic? Once we see more clearly what *our* unconscious assumptions are and perhaps where they come from, we can begin to separate our feelings from them and see them more clearly as the floating thoughts they are. So what would be heroic about this? Inevitably, we would move deeper into feeling, and feelings can be scary. Generally, our wrong and inaccurate assumptions actually prevent us from feeling our feelings by providing us with a justification:

"If I feel worthless, it's because I am worthless"

That's partly why we made them up in the first place. Ironically, while they appear to make us feel safe in our loveless hell (because at least we feel certain that we know how it really is) they actually prevent us from leaving it. Having the assumption makes the feeling make sense.

Letting go of the assumption means spinning through the vortex of not knowing anything; of not knowing which way is up. In this case:

'Do not adjust your set. Normal service will be resumed as soon as possible!'

The thing about terrible feelings is that if we are open to allowing them without trying to do anything with them, they pass. This is even true of terrible grief when a loved one is lost, although that is a perspective we do not have when we are experiencing bereavement or some kind of major separation. What prolongs the feelings and causes us problems is conflict between the feeling and the assumption. If I feel:

- Empty, separate from love, as if my heart is broken my natural response is to cry or to mourn but if I assume any of the following:
- Crying will make it worse
- It won't bring them back so there's no point
- I will never stop so I'd better not start
- I am weak to cry

The struggle to stop the crying will actually reactivate and re-energise the self blame, and drive the grief further down when what is needed most is for it to be allowed. At the end of this chapter is another exercise to take us further into an awareness of both our assumptions and attitudes.

I Know Who I Am

Allowing ourselves to feel things we have previously considered highly dangerous can be very scary at first but after living through some of the fear for a while, it becomes a relief to have let go of all that certainty. However, when we stop and think about it we notice a new set of difficulties: if we let go of our certainties where does it leave us? How can we decide what is true or not? If we are unsure whether we can really trust what we see and we have thrown our dogmas out of the window, how on earth will we protect ourselves?

In *Buffy The Vampire Slayer*, Glory, the demon Goddess, lacking any kind of cultural programming but increasingly human nonetheless, asks how people cope with raw human experience:

Glory: "How do they function here like this in the world
with all this fire runnin' through them? You have no control.
They're not even animals. They're just these meat bag slaves
to hormones and pheromones and their feelings. Hate 'em.
I mean really. Is this what the poets go on about? This?
Call me crazy but as hardcore drugs go, human emotions
are useless. People are puppets. Everyone getting jerked
around by what they're feeling..."
<u>*Buffy The Vampire Slayer*</u>

It's no small thing, the intensity of our experience when we are fully open.

Buffy, being a superheroine, knows all about this because in 'Aspect of the Demon' she becomes able to hear what everyone around her is thinking and feeling, and as you can imagine, she eventually collapses under the strain of it like an overloaded computer shorting out. Perhaps this says something about how important it is to take care of ourselves when we do this work, not doing too much at once and not putting ourselves in places with too much going on. With less certainty and more room for our own sometimes murky depths, this is one adventure we don't know the ending of. If we're to live with our fears and our frailties and learn to accept them we might need to be a little more gentle and a lot less judgmental.

Xander pretty much sums up how our weakness can become our strength when tells Buffy how badly we *all* need someone to really know us deep down – and *still* accept us.

Whether the difficult bits are revealed to us or to others, there's always the risk that we'll show the bits we are ashamed of to someone who won't like them either. Even if this doesn't happen, if we aren't going to stay the same frightened or ashamed person that we were, who will we become? When Buffy gets bitten by a demon, an aspect of its nature becomes part of her and in the same way that we become worried about the hidden parts of us emerging, she worries:

"There's this thing in me and I can't find it, I can't stop it. What if it changes me?... All of a sudden I could be something that's not me anymore."
Buffy the Vampire Slayer

It is her familiarity with and acceptance of her own fear that enables her to empathise with the experiences of others. Later, when she's recovered Buffy goes to help the guy who is in the clocktower with a rifle. She tells him that she understands his suffering but he is convinced that this is impossible because of her skill and beauty:

Buffy: "You are an idiot. My life happens to, on occasion, suck beyond the telling of it. Sometimes more than I can handle and it's not just mine. Every single person down there is ignoring your pain because they're too busy with their own. The beautiful ones, the popular ones, the guys that pick on you...everyone. If you could hear what they were feeling. The loneliness – the confusion. It looks quiet down there. It's not."
Buffy The Vampire Slayer

But Jonathan can't hear what they are feeling, only what he imagines. His own desperate loneliness has become envy and hatred as he tries to avoid his feelings of worthlessness by blaming others. If he got in touch with them directly, it might hurt for a while but then he could take care of himself and see that others might be going through something similar. It is only when we really know ourselves that we are solid enough to allow others to really be themselves. When we start to do what we need to do for ourselves, we are freed from blaming others and able to see them more clearly. In looking after ourselves, we also stop allowing and enabling others from treating us in ways that are unacceptable.

It's All Your Fault

If we have spent all our lives blaming others…

"She's pushing me around" – "He's manipulating me" –
"If I don't clean the house in the way
he expects he goes all sulky"

…what on earth do we do with our feelings and with our perceptions if we decide to stop? Our anger, our blame, our fears won't just disappear, so what's the 'right' thing to do if there is no absolute right within a partnership? What if all we really have are only feelings and beliefs from the past and in the present giving rise to and creating perceptions from our culture, from our fears, from our hopes, from our televisions? What if what we have at base is just a load of floating sensations made into meanings that mobilise, connect, magnetise, galvanise and/or inspire?

We stay with what we feel, we learn from what happens and we make choices about what we believe is right, which we modify as we learn more about what is right *for us*. We do our best, given what we know, which isn't much.

Giles: "I have volumes of law, of prophecies,
of predictions but I don't have an instruction
manual. We feel our way as we go along ..."
Buffy The Vampire Slayer

We are standing on a void, making it up as we go along with very limited information, the most important of which, by far, is our own experience. When Buffy says that she just wants to be left alone Giles asks her if that's really an option:

Giles: "You're standing at the mouth of hell."
Buffy The Vampire Slayer

The choice Buffy is faced with is the choice we are all faced with everyday: to seek truth or to use our perceptions to be bigger, louder and more right than the next person. This second way of working is endemic in our culture. Nearly everything is used to create profit. Exploitation and getting our own back has become an accepted exchange rate mechanism. This is as true within marriages and institutions as it is within the workplace. We can use others, use our position, use our knowledge to get what we want or we can use our gifts and learning to seek truth. Generally the best of intentions can go awry, but that is part of the great mystery. To stay with what we know, to acknowledge our limits, to have *good* intention, perhaps that is as good as it gets. If we do this, we are all heroes and heroines. When Buffy sets the vampire that is Angel unintentionally free, she blames herself, and Giles once again, provides the shared humanity that is perhaps our species only saving grace:

"I know that you loved him and he's proven, more than once,
that he loved you. You couldn't have known what would happen...
if it's guilt you're looking for Buffy, I'm not your man.
All you will get from me is my support and my respect."
Buffy The Vampire Slayer

Exercise

Taking care

Take a deep breath and feel how you are sitting.

Take a few moments to notice what you need to do to be more comfortable and less distracted. Ask yourself the following questions.

- Are you comfortable? Is there a part of your body that needs supporting?
- Are you hungry or thirsty?
- Are you tired?
- Is there any background noise that is disturbing you that you are able to switch off?

If the answer to any of them is yes, take a minimum of two minutes to find what you need to do to rectify your discomfort, and then take as long as you need to become more comfortable. If necessary, go for a walk and come back to this book when you want to.

Permission: 2

The last chapter had an exercise aimed at making us aware of some of the thoughts and the judgements floating around in our heads. This exercise attempts to take our awareness deeper into just what attitudes those judgements hold. You will need to retrieve the phrases from your box ready to work with them. Remember the red quality that hung around them and was put back into the box. Visualise this when you take them out.

1. Look at your phrase. How was it said? What was the attitude of the thought, or of the message? Is it said kindly or meanly? Is it loving and caring? Is it contemptuous, superior, angry? What is the attitude or feeling that is within the message? Take some time to reflect and consider until you feel you have a real sense of it. Write it down underneath the message on the piece of paper.

2. Add this new attitude to the initial message in words i.e. "*I despise myself because* I should .... and I haven't."

Allow yourself to take as long as you need to find the feeling and complete the message. Do it with only one of them at a time and only do it for as long as is comfortable.

3. Again, this is a very powerful thing to do and protecting and separating yourself from the work is the biggest priority. When you have finished, fold over the piece of paper and replace it in the box. Imagine any traces of the feelings or the thoughts as a red vapour. Imagine the last traces of this vapour pour into the box before you close it.

4. Wash your hands thoroughly. Imagine that you are washing away any of the harshness of those thoughts and feelings.

5. Remember – you are far, far more than your thoughts and feelings. Thank yourself for your hard and heroic work.

Chapter 5

Ideas To Explore

"Suffering is the one promise life keeps. If happiness comes it is a precious gift that is ours only for a brief time."
Beyond Rangoon: John Boorman

In real fairy tales, before Walt Disney got hold of them and turned them into refined sugar the endings were raw and rough and rarely involved living happily ever after. When *The Little Mermaid* gets her legs, she walks on knives and sleeps with blocks of ice at the end of her feet. After a single day of bliss with her Prince she dies *despite* the attempts of her sisters to save her. Worse, she becomes white surf on the waves because, in the laws of *their* world that's what happens when mermaids die. Normally they would live for hundreds of years, but when they *do* die they have no soul to live on as part of them.

Walt Disney was just the tip of the iceberg. These days most of us spend our lives and billions of pounds trying to avoid facing the fact that suffering and being alive go together like a horse and carriage. To us, even the words 'get real' are an attempt to avoid how much we hurt by replacing pain with a cynical acceptance that there is nothing in the universe beyond profit and loss. Where is the insight or compassion in that? We seem either to expect a romantic and womb-like safety or a cocooned high. When life catapults us from any of these into vulnerability; when death, loss, pain or illness break into our dreamworld, fear of the unknown and of being seen to not cope makes one of the most profound experiences of our lives something we try to avoid. Philip Pullman's trilogy *His Dark Materials* is about heroism precisely because Lyra and Will, the lead characters, overcome their desire to deny the full impact of what it means to be alive. When Lyra was upset, Will felt he:

'Didn't know how to comfort her,
because it was plain that she was right'
The Amber Spyglass: Philip Pullman

The idea that it is good to repress our pain as fast as possible and close ourselves to our experience is like cutting off a left arm because we're right-handed. If suffering is the one promise life keeps surely we should be on friendlier terms with it, knowing it and its own needs more intimately since it's likely to come and visit more than once.

'And at the word 'alone' Will felt a great wave of rage and despair moving outwards from a place deep within him, as if his mind were on an ocean that some profound convulsion had disturbed. All his life he'd been alone and now he must be alone again, and this infinitely precious blessing that had come to him had been taken away almost at once. He felt the wave build higher and steeper to darken the sky, he felt the crest tremble and begin to spill, he felt the great mass crashing down with the whole weight of the ocean behind it against the iron-bound coast of what had to be. And he found himself gasping and shaking and crying aloud with more anger and pain than he had ever felt in his life...But as the wave expended its force and the waters withdrew, the bleak rocks remained; there was no arguing with fate; neither his despair nor Lyra's had moved them a single inch.'
The Amber Spyglass: Philip Pullman

In the great tradition of children's fantasy novels, these stories demonstrate the reality of what life throws up. They are testaments to the way that light and dark need each other in order for there to be a story at all. The characters are weak as well as strong and actions have consequences, sometimes when the original incident is long forgotten. *The Lord of the Rings* has been a firm favourite since the 1950's with its innocence and wisdom, its trials and its triumphs. Why do we like Harry Potter so much? Because the ghastly Dursleys are so recognisable with nothing but their petty suburban concerns that it is a relief to put our finger on just how bad things can get. All of us can identify with Harry who is more human in his hopes and dreams than many of the 'muggles' or ordinary people in these stories even if he is a wizard.

It is his endless trying that rings bells for us, because this is the kind of struggle we face when we start to try and escape. When we wake up and give up addictive TV watching or heroin or narrow mindedness we welcome real life with open arms by engaging fully with the world, and it's hard work. Lyra and Will are the latest heroes but let's not forget Harry Potter, Frodo, Sam, Pippin and Merry or the Narnia books by C.S.Lewis. These tales of good and evil are reworkings of our oldest myths reflecting what is important about *our* lives as an unfolding, coherent process, as we pursue what is important and has meaning.

At this point at the start of the new millenium, we seem to be living the bit that always occurs in fairy tales where the hero or heroine is so far from home that the people she/he loves and the beliefs she/he values are a shadow, almost forgotten. Seduced like Kay in the Snow Queen or enchanted by food that is bad for us, as Edmund was by the White Witch's Turkish Delight in *The Lion, the Witch and the Wardrobe*, we watch our TV's as if we are drugged. If the programme is good we look at issues of life and death and are moved, feeling that some dim memory is stirred. Then the adverts arrive and our connection is lost as we are persuaded that the image we are being sold is more real. In C.S.Lewis's *The Silver Chair,* King Rilian is trapped underground and feels sure that the light he is looking at is merely a poor imitation of a large one called the sun, but he is persuaded that this is 'a fancy' by his jailor. How much do we live the life that has been sold to us by others for their benefit? Now, more than ever. And how do we arm ourselves against this onslaught? Only by being able to connect with what is true for us.

If this sounds a bit unnecessary, consider what happened to both *Pinnochio* (who turned into a donkey after acting like one for a few days) and Edmund, who ate so much pure sugar that he betrayed his brothers and sisters for more.

> *'At first Edmund tried to remember that it is rude to speak with one's mouth full, but soon he forgot about this and thought only of trying to shovel down as much Turkish Delight as he could, and the more he ate the more he wanted to eat, and he never asked himself why the Queen should be so inquisitive.'*
> *The Lion, the Witch and the Wardrobe:C.S.Lewis*

There are things in our lives easily as addictive as heroin and a lot more difficult to spot. Feeding the soul with sugar- be it Disney or dieting advertisements- is one way that we try to avoid suffering.

"Why can't we go to your house now?" said Edmund.
When he had first got on to the sledge he had been afraid
that she might drive away with him to some unknown
place from which he would not be able to get back; but he
had forgotten about that fear now."
The Lion, the Witch and the Wardrobe:C.S.Lewis

Unfortunately, growth food for the soul is the opposite. It necessitates awareness of the difficult bits in our experience as well as the easy.

Despair

In all of the Harry Potter books, J.K.Rowling consistently shows how Harry's fight against the Dursley's, Draco Malfoy and evil itself in the form of Voldemort (don't say that word!) is made possible by the strength of his mother's love. Lily Potter died trying to protect Harry and this left him with a sense of his own worth and existence that no amount of abuse could take away. Lily Potter's love was so powerful that it made Voldemort's death curse rebound, leaving Harry with only his legendary scar. This love is the source of Harry's power. How it works is demonstrated in the struggle with the shadowy Dementors, the terrible non-human gaolers of Azkaban. Like the Nazgul or Ringwraiths in *The Lord Of The Rings* and the Spectres in the more recent *His Dark Materials* trilogy, these creatures contaminate whatever they come near with feelings of despair:

'Familiar, icy cold penetrating his insides,
fog starting to obscure his vision'
Harry Potter and the Prisoner of Azkaban: J.K.Rowling

And what is despair but a *lack* of hope, of faith and of love?

"The Dementors affect you
worse than the others
because there are horrors
in your past that the others don't have"
Harry Potter and the Prisoner of Azkaban: J.K.Rowling

The Dementors do their work by activating past trauma. The reason why Harry is affected worse than the others is because the despair he has *already experienced* will be set off when they come near. The wound is there from losing his parents and from being attacked by Voldemort. Harry knows what fear and horror really are. The presence of the Dementors reopens this wound to convince him that horror is everything. They empty all their prisoners, sucking out their inner resources and feelings of love, comfort and hope so that their worst fears and nightmares can flourish unchallenged. In fact this is exactly the point of Room 101 from George Orwell's *1984,* one of the world's most enduring popular novels.

To really examine our experience and our feelings is to be vulnerable again. When we feel again what we felt in the past, we risk allowing those feelings to swallow us up. This is as close as we get to hell. In Azkaban, the prison for wizards and witches, this is pretty much what happens, but *because* of the absence of any redeeming love, it just gets worse and worse. Hagrid describes being falsely imprisoned there:

"Kep' going' over horrible stuff in me mind...day I got
expelled from Hogwarts...day me Dad died...yeh can'
really remember who yeh are after a while. An' yeh can'
see the point o' livin' at all...when they let me out, it was
like bein' born again, ev'ry-thin' came floodin' back."
Harry Potter and the Prisoner of Azkaban:
J.K.Rowling

There is no support, no help as Hagrid begins to hate himself and lose the sense of his own value. Ultimately the Dementors will apply the kiss of death to their prisoners, sucking out their soul until only an empty shell is left. Sound familiar?

Consider *The Lord Of The Rings* where the Nazgul had voices:

'filled with evil and horror. Ever they circled above the city, like vultures that expect their fill of doomed men's flesh. Out of sight and shot they flew and yet were ever present, and their deadly voices rent the air. More unbearable they became, not less, at each new cry. At length even the stout-hearted would fling themselves to the ground as the hidden menace passed over them, or they would stand, letting their weapons fall from nerveless hands while into their minds a blackness came, and they thought no more of war, but only of hiding and crawling and of death.'
The Lord Of The Rings:J.R.R.Tolkien

Thank goodness for Gandalf, the willing wizard whose experience has taught him not to jump to conclusions, whatever feelings you may be overwhelmed by.

"Despair is only for those who see the end beyond all doubt."
The Lord Of The Rings: J.R.R.Tolkien

If we took the experience of despair apart and looked at the belief it gives rise to, we could not get a better definition than this, for if we are convinced that the way things will unfold will be bad, if we believe that we are seeing the end 'beyond all doubt', the feeling will get stronger and be fed by our belief. The truth though is that it is only a feeling. There *is* no end beyond all doubt, only small endings within a larger awareness of life. Gandalf again:

"Whereas the light perceives the very heart of the darkness, its own secret has not been discovered, not yet."
The Lord Of The Rings: J.R.R.Tolkien

We can cut ourselves off from other possibilities, and people do, by killing themselves or their pain, but just because we close our eyes, it doesn't make it night. As a teenager struggling with terrible depression I took an overdose myself. The night before, as I was persuading myself that it was the only way, there was a tiny speck of hope but I crushed it by telling myself that I had to 'face reality': it was 'hopeless' and I ought to stop being a coward and accept how bad things really were. In my work with depressed people there has always been some glimmer poking out of the darkness, that they are misguidedly trying to extinguish but before we can really see it, we have to see what we are doing to prevent the light from growing. There is an exercise at the end of this chapter to help us really see the mechanics. This view is backed by all the best myths, which insist that the dark grew out of the light. When Sam asks Frodo whether the Orcs, which are like evil elves, eat and drink, Frodo responds:

"They eat and drink, Sam. The Shadow that bred
them can only mock, it cannot make real new things
of its own. I don't think it gave life to the Orcs,
it only ruined them and twisted them."
The Lord Of The Rings: J.R.R.Tolkien

Like the Nazgul or the Dementors, despair is really only a shadow, with no substance of its own. It can do immense damage but once its fears and concerns are fully expressed and fully heard, what was once a heavy raincloud disperses like mist in sunlight.

I wouldn't underestimate the power of despair, but I would assert that it is an illusion that feeds on itself the more we believe it to be true. It is a feeling and one place of many that we go to, but this is something we easily forget. Indeed, it is the intention of despair to make us forget:

'Here the air was still, stagnant, heavy and sound fell dead.
They walked as it were in a black vapour wrought of veritable
darkness itself that, as it was breathed, brought blindness not
only to the eyes but to the mind, so that even the memory of
colours and of forms and of any light faded out of thought. Night
always had been, and always would be, and night was all.'
The Lord Of The Rings: JRR Tolkien

Having escaped despair before, I have learned that escape is possible. Possible, but not easy. It is one thing to challenge the might of a power but another to think that just because you understand something, you can control it. I would treat it as very real, like a tiger, and not go sticking my hands in between the bars of the cage to see what will happen. I would respect it and keep my eye on it. As Gandalf says of Lord Denethor who has been having a sneaky look into a powerful seeing stone:

"He was too great to be subdued to the will of the Dark Power, he saw nonetheless only those things which that Power permitted him to see. The knowledge which he obtained was doubtless, often of service to him; yet the vision of the great might of Mordor that was shown to him fed the despair of his heart until it overthrew his mind."
The Lord Of The Rings: JRR Tolkien

Secrets

Usually the truth is simple and obvious, though not necessarily easy to absorb and act on, particularly when we are talking about our emotions. Some of the clearest descriptions of how we work can be found in the Harry Potter stories. J.K.Rowling could be talking about counselling when she points out:

'Nothing anyone said or did could make Harry feel any better because they only knew half of what was troubling him.'
Harry Potter and the Prisoner of Azkaban: J.K.Rowling

J.K.Rowling knows that you can't help solve a problem if you don't know what's wrong. Everyone knows that when we feel nauseous, we generally feel better once we've been sick so why do we find it so hard to cry when we feel a build up of grief, or express anger or even joy when we are filled with these feelings? As we've seen in previous chapters, chronic and long term inability to express ones feelings is fundamentally linked to depression and depression has its own voice. It tells us "What's the point?" and despairing of hope, we feel unable to summon any strength. We are then faced with a dammed up life source where our other feelings become stagnant. Imagine the body, where dams are caused by excess fat and waste building up in arteries and the internal organs. In the emotions there are also toxins and waste caused by disturbance, which muddies our experience and our feelings. We then develop beliefs to protect us from the pain, such as 'it's hopeless' or 'big boys don't cry' and these beliefs prevent us from grieving our loss or stating our fear or expressing our anger. These assumptions and rules are picked up from parents, schools and culture. Often the initial distress is caused by one person's disturbance being misdirected or out of control: an angry, blaming parent, a shrivelled up, bitter teacher. My guess is that we would deeply like to tell these people what to do with their pushy insistence about how things are. As adults on a path of growth, we are finding a way. As children however, it was far, far too dangerous, so we blamed ourselves and swallowed it whole, along with our anger. This is why we love Harry so much. He enacts our greatest dream and it comes true. In *Harry Potter and The Goblet of Fire,* Harry is just about to die. Voldemort has laid the Imperius Curse upon him and he is fading:

'His mother was not here to die for him this time...
he was quite unprotected... "I WON'T!" ...And these
words burst from Harry's mouth; they echoed through
the graveyard, and the dream state was lifted as suddenly
as though cold water had been thrown over him.'
Harry Potter and The Goblet of Fire: J.K.Rowling

We all need to know it can be done.

Of all the feelings my clients bring me, the one of total exposure in the face of a critical, larger force is the most difficult one to face, and the one that most dams up the flow. Recognising this, and then ignoring it, can be the way forward.

'It was easier to keep going now he had started.
It was even a relief; he felt almost as though

something poisonous was being extracted from him; it was costing him every bit of determination he had to keep talking, yet he sensed that once he had finished, he would feel better.'
Harry Potter and the Goblet Of Fire: *J.K.Rowling*

Complications

Of course, with Voldemort or Sauron or the White Witch it's not that difficult to see what we're dealing with. We are only required to be brave, but what happens when we are undermined by our inability to identify the truth? When we have no confidence in our perceptions? When the way we have been treated leaves us unable to trust anyone because we expect them to treat us badly as well? This expectation becomes our world: an experience we will look at in depth in chapters 7 and 8. It is enough here to say that if we've only seen power used to dominate, exploit and control us, the words 'abuse' and 'power' become stuck together in our minds, as if they are one concept. The truth is that having power is not the same as abusing power. It can mean having the power to step into the shoes of those whose failings you can't understand so that you *can* explore and understand them. Why should someone with power *necessarily* abuse that power? Power *in itself* is neutral; it is how power is used that is important. Perhaps one of the worst experiences we can have is when power is used to destroy the confidence in our own perceptions. In England the notion that God can see every little thing you're thinking has just such an effect, nipping any 'unacceptable' views or feelings in the bud before they are fully formed. Often this means our knowing becomes a taboo instead of a route to wisdom. My Grandmother told me God had a black book that he put black marks in and of course because he was God, all seeing and all-powerful, I couldn't get away with anything, and that included freedom of thought!

"And thy shrivelled mind be left naked to the lidless eye."
The Lord Of The Rings: *J.R.R.Tolkien*

Somehow we have to build confidence in our felt experience so that we can really assess what is good for us. Then we can connect with what it is about others and about ourselves that we struggle with. As Dumbledore, headmaster at Hogwarts says:

"Understanding is the first step to acceptance and only with acceptance can there be recovery."
Harry Potter and the Goblet of Fire: JKRowling.

Knowing our feeling and allowing it opens the road for compassion, for shared humanity, forgiveness and for shrewd assessment. Recently I had a client who I had praised for being 'sensible' but our explorations showed that to her 'sensible' meant rigid and critical and dismissive because of the way her parents had used the word. It's not a word I now use with her and it is a word that she is aware holds a lot of restrictive messages and attitudes, just to *her* because of the way it *was* used which gave a very different feeling to the one I intended.

Flipsides

'Pain is love turned inside out and stretched across the barrel of time'
Between Psychology and Psychotherapy: Miller Mair

Everything true has a flipside. Without suffering we would see no happiness, without the dark we would see no light, without grief we would not experience joy, without madness there would be no sanity, without restriction there would be no freedom. Without opposites, dualism and contrast, we would not be aware of anything. There would be no consciousness and no humanity. If only we could embrace that these opposites are central to and inseparable from our existence we could live life to the full. On returning to the Shire and finding it exploited with its inhabitants sold into slavery, the hobbits are heartbroken, but even they recognise that, as Sam Gamgee says:

"It comes home to you, as they say; because it is home and you remember it before it was all ruined."
The Lord Of The Rings: J.R.R.Tolkien

The opposites in life are an essential. I assert this not because I believe it but because I have experienced it to be true. Yet the irony is that before I had done this, I didn't know it to be true. Transforming our pain seems necessarily to be about embracing it while doubting that we will ever escape. This takes a certain amount of active faith. Any theologian will tell you that faith is the flipside of doubt and vice versa. And in the process of counselling, faith and doubt stand side by side.

The meaning provided by these opposites and contradictions seems to be larger than our small perspective on life can grasp but it does, at least, leave us with a sense of wonder and openness to what will happen next! There is of course pain, but as one of my clients noticed recently: if she hadn't been forced into looking after herself by being so neglected, she would never have developed the hardiness that enabled her to travel the world fearlessly. For Sally, the flipside of her neglect is a capacity to be independent at a level far greater than that of her peers. Another client can see clearly that although her feelings of panic and self hatred are telling her not to come back to continue her counselling, the stubborn streak that she developed as a child in order to stop herself being completely pushed around will enable her to return next week even when she doesn't want to. For myself, the flipside of my internalised message about being useless is that I have three degrees and am writing this book. OK, I'm driven by a negative message but I cannot ignore that I have developed the skills to achieve a lot. Like it or not, we are endlessly tied to our history, which plays an irrevocable part in who we become.

A few of the laws to do with flipsides are simply this:

- Anything with the power to do good can do harm in the hands of humans
- If you take on a role that protects you it will also restrict you
- What draws you to someone is also what drives you mad
- Ambivalent or opposing feelings in close relationships are part of the deal and must be embraced and worked with

If we allow ourselves to realise our history and our feelings fully, we are able to transmute that energy into something we can use for ourselves instead of it using us. In the long term, impossible as it sounds from the short-term perspective, we could even say that we benefit from, as well as being damaged by, our abuse. In the long term it is what we do with our scars and our skills that matters more than how we got them. As the wise and compassionate Dumbledore says:

"You fail to recognise that it matters not what someone is born but what they grow to be."
Harry Potter and The Goblet of Fire: J.K.Rowling

Exercise

Permission: 3

Whose voice?

For this exercise, as for the other ones you will need a safe place where you will not be observed or interrrupted. You will also need a couple of chairs. By this time you should have a message or several messages in your box that contain a command, or perhaps a judgement about yourself and an attitude i.e.:

"I must set the video. I'm so forgetful and useless. I'm not worth anything really"

or

"I shouldn't have shouted at my son. I'm an impatient intolerant mother and I'm so selfish I'm pathetic.

1. Choose one message and change the 'I's to 'You's' so that it reads 'You must set the video. You're so forgetful'

2. Get your chair and imagine yourself seated in it. Then stand across from it. Say your phrase or phrases out loud to your imaginary self.

3. Move around the room. Sit down opposite it. Go upstairs and shout the message down. Try and get a sense of what is most familiar, of what 'fits' with your experience and stay with the position that works best.

4. Whose voice is it? Perhaps it was your mother's or your father's, or maybe one of your sibling's.

5. Repeat this with other messages if you feel you have the time at the moment. Write the name of the person whose voice it is on the paper and place it back in the box. Take as much care as before to wash away any violence or harshness, imagining these feelings as red and seeing them firmly back into the box. Wash your hands and acknowledge your bravery.

* * *

This is deep and difficult work so it is essential that you do it at a pace that you are comfortable with. If you have got this far, you may be able to take the next step or you may not. If you can, it should simplify things a bit. If not, this should not prevent the exercise from helping. Some people (not everyone) will be able to see a sort of 'common denominator' belief to these messages: a single thought or belief, which underlies all the others – which is connected to all the others or which the others 'spring out of'.

What is the common denominator that covers all the phrases you have considered? There may be more than one. It is probably more general than the phrases you have. Take a pen and paper and write down the messages that you remember. Next, allow yourself to sit with these messages. Watch your mind and where the message leads it. See if you can catch the bottom line belief about yourself as you track your judgements or messages.

Here is list of the assumptions that appear regularly in this book. Perhaps some of them fit your own:

- I am a very bad person.
- I am worthless.
- I don't even exist.

However – this can not be forced. Awareness of self often pops up when you are least expecting it – perhaps when you are relaxing or even dreaming. Now that you have started this process it will begin to run itself because of your focus. If you do get hold of one though, put it into the box and ritually clean yourself to separate from it as well as thanking yourself for your efforts.

We are getting very close to core assumptions. Often we are so depressed or distracted we don't even know they are there. They are entrenched precisely because they do form part of our core. They are our image of ourselves but they come from someone else's view of us. They are not all that we are and they are certainly not fixed.

Chapter 6

Ideas To Embrace

One of the things that make us love Harry Potter so much is that he is just as subject to the workings of the universe as we are. Magic has to be learned, generating exactly the same struggles and mistakes that chemistry, maths or food technology hold for the average schoolchild. When Mrs Weasley cooks, she doesn't just wave her wand to instantly summon up a ten course meal, she has books and a recipe. They have *systems* of magic as much subject to the laws of what is possible and impossible as our own systems are. This is what makes Harry real, rather than ridiculous. Hogwarts, the school for wizards, is fresh and funny, but it is also familiar. The pictures on the walls may be full of animated people, but like the people in our world they moan or are vain, a ghost might be unusual but its personality disorder is not entirely from another world and there's always a teacher who hates you! Harry is as much a victim of unfairness and discrimination as we are. It's just that he has some magical powers, which are very apparent. Much more apparent than knowledge of the magical laws in our own world. Yet just about everything that happens in Harry Potter that is deeper magic, <u>not</u> text book magic, happens in our world too, but it doesn't mean for either us or him, that this is easy:

"Did you think you could snap your fingers, and have it as a gift?
What is worth having is worth working for"
<u>The Amber Spyglass</u>: Philip Pullman

The laws of *human* reality are just as magical, and just as subject to restrictions in our own world. They're just not very well known.

Connections

In our culture we tend to believe that the terrible feelings referred to in the last chapter are dreamt up but as also mentioned, the nightmare feelings we are trying to avoid have often *already been experienced by people* and lay buried because they are so awful. Ironically then, we find ourselves trying to shut the stable door after the horse has bolted and, perhaps because our traumas are constantly trying to be accepted and added to the sum of our experience and wisdom, our traumas are constantly set off by relationships, images in the media or stories that we identify with. Because we are so literal, and have so many fears about 'indulging' instead of embracing these experiences, we tell ourselves not to be ridiculous. Unfortunately, at the same time we prevent ourselves from growing. In *The Amber Spyglass* it becomes apparent that 'Dust' is really human awareness but as Xaphenia warns:

> *"Dust is not a constant. There's not a fixed quantity that has always been the same. Conscious beings make Dust – they renew it all the time, by thinking and feeling and reflecting, by gaining wisdom and passing it on."*
> The Amber Spyglass: *Philip Pullman*

By not feeling and reflecting, we cut ourselves off from our souls and with our souls go our deeper connections to other human beings.

The Killing Fields was a film that particularly affected me. I cried for a week, yet what possible connection could I have with the terrible genocide and dictatorship of Cambodia in the 1970's? When I began to trace what I found so upsetting, what I saw and felt, I reached a core of terror and distress all of my own and found where it fitted in my life. Towards the end of this film, the central character who has escaped from a work camp against unbelievable odds is asked by someone to try and take his son across the border to safety. Now not only has he an impossible task ahead, but he makes it more difficult by taking on this young boy of about 3. He carries him through miles and miles of mined and almost impenetrable jungle, frequently hiding from Pol Pot's soldiers. Very close to the border, exhausted, this child clinging to him, his next step produces a sudden click and you know and he knows it is a mine. When he steps away, it will blow up beneath him. He has to do it fast. He has no choice. He does not know what will happen. He moves. It blows. The child dies but not in a sensationalist way. The man looks at the child to see if it has survived the blast and it simply hasn't. It has been hit. There is a very small amount of blood. The child is dead.

What happens next is the bit that makes me most hysterical. The man just shakes his head and drops it down. He is completely defeated. He has tried so hard. He has done his absolute best and it just isn't good enough. He has exhausted himself by being on the lookout for both of them. He is starving, wounded, burdened and yet he still persevered, but no matter how hard he tried, how much he knew, *he didn't know where the mines were, so he couldn't avoid them.*

In my own, outwardly safe, aspiring-to-middle class home, I had my own inner experience of life and death, of despair and defeat. My father was emotionally volatile and could destroy hopes and realities with words, and because he *was* emotionally volatile I never knew where the emotional landmines were and I was always stepping on them, even when I tried really, really hard not to. Since they were there and it wasn't my fault there was absolutely nothing I could do about it. To take it even further back, I was already predisposed to be drastically affected by his bullying because of an even earlier experience of losing my mother due to her near death with kidney problems at an age where I couldn't possibly make sense of that. Screaming and screaming for the warm breast that you have suddenly lost at 3 months and not being comforted taught me that loss leads to trying to redress that loss which ends in the silence of despair. The film character's hopeful/hopeless struggle connected me absolutely with those feelings and how they had come about and also how they were linked together in my mind.*

As I write this I am already judging myself for using something as significant as the genocide in Cambodia to make me aware of my own experience. Next to wars and the death of children, it seems petty even to mention my own pain, yet I know there is no competition. Pain is pain. The point is that our inner experience has to find a voice and a home if it is to stop incapacitating us. It is our deep connections that make us able to reach out to others and being rooted in our own experience that allows us to accept and to allow the differences of others. We have no way of comparing or grading our feelings. They just are, yet these judgements often prevent us from fully examining and reflecting on them.

Dumbledore has no such reservations. In fact, being a wizard, he has a wizard way of doing this:

*The value of physical warmth and contact is paramount at this age. Harlow in the appendix

'Dumbledore drew his wand out of the inside of his robes and placed the tip onto his own silver hair, near his temple. When he took the wand away, hair seemed to be clinging to it – but then Harry saw that it was in fact a glistening strand of the same strange, silvery white substance that filled the Penseive. Dumbledore added this fresh thought to the basin.'
Harry Potter and the Goblet of Fire:J.K.Rowling

The basin is called a 'penseive' and provides a container for all that Dumbledore knows and has experienced:

"I simply have too many thoughts and memories crammed into my mind...one simply siphons off the excess thoughts ... pours them into the basin and examines them at one's leisure. It becomes easier to spot patterns and links, you understand, when they are in this form..."
Harry Potter and the Goblet of fire: J.K.Rowling

Sounds as if it has remarkable similarities with the counselling room to me!

Vulnerability

One of the things that defines us as most human is that we never know what might happen. Change is most definitely going to arrive, however long we spend trying to avoid it. The challenge is to find a way to embrace it.

In C.S.Lewis' novel *Prince Caspian*, Lucy, Susan, Edmund and Peter return to Narnia to help the rightful King but they are hampered in their journey by the unfamiliar landscape which has changed dramatically since their last visit (hundreds of years have passed). Because of this, they get completely lost. When Lucy, the youngest, sees Aslan telling them to go the other way, she tells the others, but because they cannot see him, and she is the youngest, they ignore her. That night Lucy wakes and sees Aslan again. After a reunion filled with bliss (for Aslan is her God), he says:

"We must not lie here for long. You have work in hand and much time has been lost today"
"Yes, wasn't it a shame?" said Lucy. "I saw you all right. They wouldn't believe me. They're all so..."
From somewhere deep inside Aslan's body there came the faintest suggestion of a growl.
"I'm sorry," said Lucy... "but it wasn't my fault anyway was it?"
The Lion looked straight into her eyes.
"Oh Aslan.... I couldn't have left the others and come up to you alone, how could I? ... You mean... that it would have turned out all right somehow? But how? Please, Aslan! Am I not to know?"
"To know what would have happened, child?" said Aslan. "No. Nobody is ever told that."
Prince Caspian: C.S.Lewis

We are all vulnerable because we don't know the outcome of our choices. Lucy didn't dare go against her brothers and sisters, they arrived at the battle late and several close friends had been killed. She will never know if that would have been different had she followed Aslan earlier. Our strength lies in acknowledging that vulnerability. Not knowing the right course of action, all we can do is our best, make our choices and live our lives true to our beliefs. Even when dying, Theoden, King of the Mark of Rohan in Tolkien's *Lord of the Rings*, has triumphed in his spirit, having turned at the last from despair and distrust to follow his beliefs:

"I go now to my fathers. And even in their mighty company I shall now not be ashamed. I felled the black serpent. A grim morn and a glad day and a golden sunset."
The Lord Of The Rings: J.R.R.Tolkien

Only in knowing that our vulnerability has to be embraced will we be on the right track to living well. In C.S.Lewis' *The Voyage of The Dawn Treader*, Eustace (cousin to Lucy and the others) has been a victim all his life of his parents telling him to be superior. Consequently he has built up layers and layers of pretence and defence against feeling vulnerable to the point where he no longer knows who he is. Instead of making friends with his cousins and sharing the difficulties, he spends his time trying to look better than everyone else. Eventually, on an island visit he sleeps on a dragon's hoard of treasure, and:

'Sleeping on a dragon's hoard with greedy, dragonish thoughts in his heart, he had become a dragon himself.'
The Voyage of the Dawn Treader: C.S.Lewis

Eustace has no idea how to behave until he starts to suffer and I blame the parents! As the narrator points out:

'Most of us know what we should expect to find in a dragon's lair but, as I said before, Eustace had only read the wrong books. They had a lot to say about exports and imports and governments and drains, but they were weak on dragons.'
The Voyage of the Dawn Treader:C.S.Lewis

Eustace is not well equipped in understanding the laws of life at all. Eventually Aslan comes to change him back. He takes him to a well for bathing but tells him that he must 'undress' first. Eustace realises that he is scaly, like a dragon, and scratches off some scales, but it is not enough. Again Aslan tells him to undress and again he tries, but he remains a dragon. Psychologically speaking, the rigid layers of belief he has adopted as emotional armour have become a prison. He no longer knows how to feel and has spent so much time building up false ideas that they will hurt to remove, like ripping off a plaster, which of course, only our mothers could ever really face doing, because they knew it was for the best. Eustace describes it, later, to his cousins:

"The very first tear he made was so deep that I thought it had gone right into my heart. And when he began pulling the skin off, it hurt worse than anything I've ever felt. The only thing that made me able to bear it was just the pleasure of feeling the stuff peel off...Then he caught hold of me – I didn't like that much for I was very tender underneath now that I'd no skin on – and threw me into the water. It smarted like anything, but only for a moment. After that it became perfectly delicious..."
The Voyage of the Dawn Treader: *C.S.Lewis*

The stronger our emotional armour, the more difficult it is to get off. The more we feel our pain, the less frightening it becomes and the faster it flows through us. The other choices don't really seem to be worth having. We should be less sensitive, I hear people say, but my own experience is that I simply am this sensitive and, if you choose to judge it this way, I am this pathetic. If befriending your inner patheticness is the only way to embrace the mystery of the divine grief process, so be it. It's very, very difficult and our resistance and our anger in the face of the pain of connection and loss is perfectly understandable. Perhaps Giles speaks for all of us when he says of Angel:

"He's likely to strike out at the things that make him most human."
Buffy The Vampire Slayer

So how on earth can we bear it at all?

Love

Our stories and songs are filled with images of the power of love. Who can ever forget Tinkerbell, poisoned by Captain Hook, choking as Peter Pan wills us to will her back to life:

"If you believe in fairies, clap your hands."
J.M.Barrie: Peter Pan

ET does it, Aslan does it, Harry Potter does it:

"The Patronus Charm (acts as) a shield between you and the Dementor. (It) will work only if you are concentrating with all your might, on a single, very happy memory"
Harry Potter and the Prisoner of Azkaban:*J.K.Rowling*

Despite the pain of replaying his parents last moments in his head, Harry needs to reconnect with a time when they were alive and, sure enough, the love they bore him protects him against despair, just as the mothers love for the infant is essential for the developing sense of self.

"You think the dead we have loved ever truly leave us? You think that we don't recall them more clearly than ever in times of great trouble? Your father is alive in you, Harry, and shows himself most plainly when you have need of him. How else could you produce that particular Patronus?"
Harry Potter and the Prisoner of Azkaban: *J.K.Rowling*

This is a magic so familiar that we are in danger of undervaluing it. In the counselling room, strength and belief in the authentic self is something accessed by creating conditions where love and trust in the self can be experienced. Like all real magic, its source is a mystery, but like mothers, counsellors can be the instruments through which love can find a way to be felt. Harry Potter knew what it was to be loved so much he couldn't fail to feel worth something:

"His mother died in the attempt to save him – and unwittingly provided him with a protection I admit I had not foreseen...I could not touch the boy...His mother left upon him the traces of her sacrifice... this is old magic.... My course was deflected by the woman's foolish sacrifice, and it rebounded upon me."
Harry Potter and the Goblet of Fire: *J.K.Rowling*

Older magic than any of Mr Weasley's spells to make the car fly. It would seem that love is a force all its own. We can't buy it or steal it or will it or control it. This is why Voldemort, like the Narnian witch, can never win, but can only surrender. As Voldemort also says:

> *"I wanted the blood of the one who had stripped me*
> *of power 13 years ago, for the lingering protection his*
> *mother once gave him, would then reside in my veins too."*
> *Harry Potter and the Goblet of Fire: J.K.Rowling*

But killing Harry Potter will only lead Voldemort to despair over what it is he has never really had. For the Witch, again, the immortality she takes by force by eating the apple of knowledge can only stretch life into an endless torment, because by taking something this way she only reinforces her own absence of love, or hope, or change and therefore, as Aslan says:

> *"That which is joy and life and health to you,*
> *is death and horror and despair to her"*
> *The Magician's Nephew: C.S.Lewis*

Cycles

If so much of our growth is about being vulnerable, and much of that seems to involve going backwards into past feelings, then how on earth do we move forward? Perhaps if we begin by recognising that our patterns of behaviour are learned and simply repeat a way of being without really thinking about it, it will be easier to see that we are largely standing still. How often have any mothers out there noticed that they have done or said something which makes them feel as if they are turning into *their* mother? We live in a world of high-speed change. Some of what was deemed acceptable in our parent's youth is illegal now and, more than ever, questioned or challenged by freethinking individuals anyway. At least sixty percent of my clients come to me with more sympathy for their parents than is good for them, by which I mean that they are more familiar with the perspective of their parents than they are with their own. As a consequence, they often cannot tell what they feel because of the muddle, as if parents could hear our thoughts or criticising the parental view would be a betrayal that would end in death and destruction. Perhaps it would have done if the client were still an entirely dependent baby. Perhaps that's partly why it's so frightening.

Sometimes it's like we really have become our parents. We were so small and they were so big. The bigger they were and the louder and the less they really saw and encouraged us, the more we swallowed them and their beliefs whole. The momentum of the generations stacked up behind us is so powerful that whenever I see what I am trying *not* to be like, my task seems to be akin to stopping the world on its axis and pushing it back the other way. The blaming of those outside the family and those inside, the dumping of anger and guilt and grief and unhappiness onto someone else to express in its raw form; the scapegoating and stereotyping of individuals until opinions about their characters seem like incontrovertible fact is common practice. In many families holding a stiff upper lip is considered the highest of achievements and the one who doesn't hold it is considered the 'unstable' one. It's not easy to walk the line of saying it how it is and not falling into repeating the blaming either. Philip Larkin may have been right when he said:

'They fuck you up, your mum and dad'
Philip Larkin: This Be The Verse

But that doesn't mean that they meant to, any more than I meant to fuck up my own kids.

So, sorting it out, challenging the momentum of centuries and changing the future is easily as heroic a task as Frodo's taking the Ring of power to Mount Doom in Mordor, and about as difficult to deal with, but once on the path, it becomes obvious that there's no going back. As Sam Gamgee says:

"Of course he's had a bit of schooling, so to speak –
we all have – since we left home, or he'd be so terrified
he'd just fling the Ring in the river and bolt."
The Lord of The Rings: J.R.R.Tolkien

They live within us, our parents, for good or for bad, there to call on when we need them or calling on us when we need them least. After all those years of input, we know what they would have said. For Harry Potter, when he is in the greatest danger, knowledge of his father's strength and power come to enable him to get rid of the Dementors with his Patronus charm:

"I saw me but I thought I was my Dad."
<u>*Harry Potter and the Prisoner of Azkaban:*</u> *J.K.Rowling*

There's a lot about Dads. Poor Dads, poor Mums. Doomed to fall short of their greatest hopes for themselves in that role! But it doesn't matter. My father really went for the Philip Larkin school of truth, but weeks after his death as I cried and cried, it was him who eventually stopped me, sitting on my bed in his blue dressing gown, stroking my head, shushing me and saying that that was enough – giving me, in this instance, a gentle permission to stop. For Harry it's Prongs he sees, Harry's father in the form of a stag. True to life, Bambi was the Disney film I ran out of screaming as a child. His mother gets shot, and eventually, like the Lion King, Bambi stands proud and alone, adult at last. He has put away childish things. He has faced his truth and that was all he could do, but he has moved forward.

Growth

So our parents are inside us, for good *and* bad. I've said a lot about the good, but what about the bad? Recently I have had clients who have noticed how much pain they are in because of the way their own thoughts attack them. If something goes wrong or they are upset, they say it is because 'I'm so stupid', yet if we have higher regard for ourselves we tend to say 'never mind' to ourselves. In this age of cancer, where cells multiply and eat the goodness in their victims, or AIDS where the immune system just can't fight back we seem to have an equivalent psychological *immune deficiency syndrome*. We may know little about controlling physical diseases with our minds but in the field of psychology, recognising what we are doing with our minds makes changing those habits quite possible with the right support, but how big a sense of ourselves is enough to make change possible?

"I think the only reason I never lost my mind is that I knew I was innocent...it kept me sane and knowing who I am...the Dementors couldn't destroy it...it was an obsession...but it gave me strength."
Harry Potter and the Prisoner of Azkaban: J.K.Rowling

How did Padfoot/Sirius know this? Because he knew what the truth was and he was strong enough to hang onto it through thick and thin. As Nietzsche, the great German philosopher said:

"That which does not kill me makes me stronger."

But strength and power can be for both good and evil. What makes some people embark on a path of hatred and destruction and others spend a lifetime trying to overcome them? What is 'good enough'?

'To crooked eyes truth may wear a wry face.'
Lord Of The Rings: JRRTolkien

We often see what we expect to see and self-hatred breeds a habit of seeing the darkest motives in others. The less aware we are of this voice and where it comes from, the more powerful it is. The delusion is that we may think we are using it to protect ourselves but the truth, as Gandalf the holder of laws knows, is that:

"Perilous to us all are the devices of an art deeper than we possess ourselves."
Lord Of The Rings: J.R.R.Tolkien

Self-acceptance gives us a much lighter and more optimistic perspective. Dumbledore is the father we would all want because his acceptance and understanding makes all his charges (especially Harry) know they are 'good enough'. Consequently they often feel the same about others. Both Moody and Voldemort were known to have had *'very disappointing fathers'* and see nastiness, incompetence and the lowest of motives or achievements in everything. When Harry loses the race trying to save the French contestant, the poles apart perspectives of Voldemort and Dumbledore, or even good and evil are illustrated when Voldemort says to Harry:

"Luckily (for me) Dumbledore took your
idiocy for nobility and marked you high for it."
Harry Potter and the Goblet of Fire: JKRowling

Ultimately we can all *know* the impact of feeling 'good enough' as people, and therefore we can choose to bestow it on others. This choice – to try and live by the values that are helpful – is one outcome made possible through our knowledge of psychology. The alternative, the unconscious belief that we are 'not good enough', limits our choice and the direction of our growth. Our resentment and our pain and our beliefs about ourselves are very significant and need attention or servicing, like a car, but the direction we take is determined by our values and what we believe is right so we have a real responsibility to work out what they are and hold them consciously. For better or worse, conscious or unconscious, they make a difference.

Paradox

As I struggle trying to find definitive answers to these questions of how much sense of self do we need? How much good is enough? I notice that I am armed with lots of information about the nature of things and I still don't know. Perhaps I am looking at it all the wrong way with the wrong question. Instead of wanting the meaning of life perhaps we should be asking what is the *knack* to life? Instead of treating life as if it is a straight road with a destination at the end, we should treat it as an endless layering of concentric circles and spirals rather like DNA. Just like Mary Malone in *The Amber Spyglass*, if we want to see something important perhaps we have to look at it sideways and wait for the answer to appear. If we get half an answer, we should treat it as a clue; consider what it means to us. The more we try to grasp something tightly and *completely*, the more difficult it will be to get hold of. This and the following statements are all paradoxes. Although they are contradictory and appear to be absurd, they do in fact *tend* to be true:

- If we give up trying so hard to be worth something we feel better.
- If we fully experience our infantile feelings while holding compassion for ourselves we are likely to grow up.
- The more I know who *I* am, the less I know about anything.

This last piece of paradoxical wisdom came from Socrates and is echoed by Dumbledore. Dumbledore is able to acknowledge his limits, humbling Harry and us with this seemingly absurd truth. How can a wizard who is so powerful not save the world?

> *"I have no power to make other men see the truth or to overule the Minister for Magic"...Harry...felt as though the ground beneath him was falling sharply away. He had grown used to the idea that Dumbledore could solve anything. He had expected Dumbledore to pull some solution out of the air."*
> *Harry Potter and the Prisoner of Azkaban: J.K.Rowling*

Magic – that is, deep magic – is paradoxical rather than static with a fixed answer. Nowhere do we see this more beautifully illustrated than in the contrast between levels of wisdom and magical ability. We are gripped by Harry's aptitude at broom flying or Charlie and Fred Weasley's exploding sweets but these are craft rather than art. As I have said already, they can be learned like Mrs Weasley has learned to cook from her magic, as opposed to non-magic cookbook. What's the difference between a spell and a recipe? Eventually a spell saves time and takes less effort, but then so does a Kenwood Chef or a vacuum cleaner. We understand the Dark Arts lessons because it is *like* non-magic learning: like learning to drive a car or make a bomb or bake a cake. It is using new and different tools but it is in the same order of things. In *The Magician's Nephew*, the Witch demands to be taken to the Magician who has called her into the world. When all she meets is cowardly weak Uncle Andrew she says:

"I see, ... you are a Magician of a sort. Stand up, dog, and don't sprawl there as if you were speaking to your equals... I see what you are. You are a little, peddling Magician who works by rules and books. There is no real magic in your blood and heart."
The Magician's Nephew: *C.S.Lewis*

Real magic lies in understanding the laws of reality, but it's trusting them that really counts. In paradoxical, or very deep magic, there are always two parts for real transformation to occur. In *The Lion, The Witch and The Wardrobe,* Edmund is enchanted/seduced by the Witch and her Turkish Delight and turns away from what matters: his family and what's right, betraying their whereabouts to the Witch. Because he has turned bad, *he* belongs to the bad, in this case the Witch. It is simply the way of things. Deep magic from the dawn of time *is* the way of things.

"Fool," said the Witch...do you really think your master can rob me of my rights by mere force? He knows the deep magic better than that. He knows that unless I have blood as the Law says, all Narnia will be overturned and perish in fire and water."
The Lion, The Witch and The Wardrobe: *C.S.Lewis*

In a chapter that still makes me shiver, Lewis points to a still deeper magic. A paradoxical magic that illustrates the greater power of light, faith and trust. The Witch knows that if you betray someone you move into the darkness; that you blind yourself to the pain you've caused and move further from yourself. You blind your '*self*'. When she says that Edmund 'belongs' to her, she's not arguing about it. She is simply stating what is. *It is not possible* to change him with a magic wand, or force. It is *not possible* to buy him. You cannot buy truth or true human acts of the soul. If you could, there would be no order, and the world, having lost its larger laws, would dissolve into chaos. This is deep magic, but it is also straight line thinking. It's obvious. I can't make black become white or white become black. If something belongs to me it's mine, not yours, and even if you take it, it doesn't make it yours. It only becomes yours if I give it to you wholeheartedly without knowing what you will do with it. This introduces a whole new dimension, another element where rather than *making* things happen, and *keeping* control, I *stop* trying and I *let go of control of the outcome.*

In the chapter 'Deeper magic from the dawn of time', this is what happens and it is important to recognise that at this point, Edmund has recognised how wrong he has been, how he wishes he had not betrayed the others. The old way of being has been superseded by the introduction of a new element, a letting go. This new element constitutes a major difference and what happens next shows that what is apparent is not necessarily true. If, for instance, I love my daughter and want her to come to no harm, the obvious thing to do would be to lock her up, keep her away from men or not let her drive a car. Paradoxically, in giving her freedom instead, she learns to trust herself and me and is more likely to talk to me openly and listen to what I have to say. She is therefore in many ways, safer, but I have to *trust* that this will work. I have to *give* her that *trust* in a real act. I cannot *pretend* to trust and keep her locked up. I have to *really* do it. I do not *know* the outcome. I have a hunch about it. I *believe* it. The older I get and the more I do it the more I know it to be true, but this is surely the position we are all in when we face letting go of our familiar ways of reacting and move into our true, innocent selves. We fear that we will die, but in fact we are redeemed.

'Though the Witch knew the Deep Magic,
there is a magic deeper still which she did not know...'
The Lion, The Witch and The Wardrobe: C.S.Lewis

The magic that the Witch does not know is the magic outside of her blind lust for power, revenge and immortality. The power of love, which cannot be won, only surrendered to.

"If she could have looked a little further back, into the stillness and the darkness before Time dawned, she would have read there a different incantation. She would have known that when a willing victim who had committed no

> *treachery was killed in a traitor's stead, the Table would crack and Death itself would start working backwards."*
> The Lion, The Witch and The Wardrobe: *C.S.Lewis*

The 'incantation' is the beginning of all things. It is the love that will catch us if only we are willing to let go and fall. While this may be how paradox works theoretically, those reading may be more familiar with how it works on a practical level. In straight line thinking, we all believe that if we let go into our grief we will fall into a bottomless abyss which will never end. When Aslan, the willing victim and the incarnation of love, is killed, this is what happens to the watching girls:

> *'I hope no one who reads this book has been quite as miserable as Susan and Lucy were that night; but if you have been – if you've been up all night and cried till you have no more tears left in you – you will know that there comes in the end a sort of quietness. You feel as if nothing is ever going to happen again. At any rate that was how it felt to these two.'*
> The Lion, The Witch and The Wardrobe: *C.S.Lewis*

In the New Testament, Jesus was constantly indicating that he who was first would be last, that it was easier for a camel to go through the eye of a needle than a rich man to get into heaven. The story at the heart of the Christian myth is a paradox, the same paradox that is at the heart of all things Pagan: that if you 'die to yourself' – that is, if you let go of all that you *think* you are, all the certainties about yourself, you will be reborn. Some of the more evangelical parts of the church would argue that it is essential to recognise that someone else died for you, so you don't have to. That is, that Jesus did the sacrificing already, but I think this is just a ruse to get people connected to actually how powerless they really are, which is, of course, part of the process. Unfortunately, the church just as often convinces people that they don't have to do anything except acknowledge they are sorry and ask Jesus to sort it out which, judging by Eustace's experience as a dragon, isn't enough. We need to have faith in the love, and if people can do this through seeing

Jesus's love as so great that he sacrificed himself for us, fair enough. Whatever gets to the heart of the matter.

The process of dying, dormancy and transformation is all around us. It is only the perspective of our everyday-in-the-world self, or of our determined-to-cling-to-what-we-know self that makes us scared, and, paradoxically, we have to live through the terror before we really know that there is nothing to actually be afraid of. For me, Jesus is a profound metaphor representing the dynamic process of life. His story indicates that our bleakest and most difficult moments happen just before the dawn (when most people die) and that whether we believe in it or not, things always change. It is simply not in the nature of things for us to be able to skip the difficult bits and still get the experience we require to learn this stuff.

In relation to pain the paradox is that when you go into it, it shifts. The paradox of therapy is that people come to it to feel better but the therapist's job is often to connect that person to their felt sense, like a doctor who presses a sore spot and asks, "does this hurt?" Clients will often use this as a reason not to come back, but if they don't come back they often remain stuck in the pain instead of shifting it. Of course, clients always expect their therapists to be very well balanced and healthy people. This, again, is straight-line thinking. The paradox is that other wounded people have better hunches about what's going on and what will help, because they know about the pain, the unhealthy thinking and the avoidance tactics. That's how they know which spots to press and when pain is being sidestepped.

Embracing paradox is never easy. We like to know the whole truth but the paradoxical thing about truth is that as soon as you pin it down it crystallises into dogma. Truth is alive and always changing. I said paradox has two parts, but we, with our limited understanding are only one of them. Even if ultimate truth could be bought, we couldn't receive it and anyway, truth is like love. We seem to be only able to stand so much of it yet we really need it to fully see and feel the profound nature of our experience. At the end of *Harry Potter and the Goblet of Fire*, Mrs Weasley's love is all redeeming:

'He had no memory of ever being hugged like this, as though by a mother. The full weight of everything he had seen that night seemed to fall in upon him as Mrs Weasley held him to her.'
Harry Potter and the Goblet of Fire: J.K.Rowling

Far from them all living happily ever after without a care in the world, Harry is moving into his feelings.

Exercise

Permission: 4

If you have done the previous exercise you will be in touch with some of the most basic messages around which you behave, believe and live most of your life. They form the reference points telling us who we are. Again, you will need to be safe and not at risk of being disturbed. You will also need your chairs, a pen and paper and a message/ messages or core assumptions from your box.

1. Sit yourself down on the chair. Imagine leaving yourself there. Step away. Now direct your core assumptions or other message towards the 'you' left behind in the chair.

2. Sit back in the chair again. How does it feel to be hearing those things? Try to put your finger on the feeling – perhaps you feel afraid, ashamed, stubborn, defiant, angry? Take the time to find the word that fits *your* feeling.

3. Step away from the chair and imagine a child that you love experiencing the same thing. A child you were committed to caring for or nurturing. It may be your own child. How do you feel towards them? My guess is that you feel very differently than you do towards yourself.

4. Write about this child considering the following questions:

- What does he/she need to feel better?
- What does he/she need to feel safer?
- What kind of different messages does he/she need to build a self-image of good enough as opposed to not good enough?
- Are these messages realistic? Is what you are saying about them true bearing in mind that they are children who have little responsibility and are very vulnerable by virtue of their age?

By now, I hope that you can see how profound these messages are and that the healing lies in kindness and care. For transformation to be possible we need to 'tune in' to our needs – to what is too much or enough, and to 'clean up' the remnants of our psychic work by being aware of its power.

Make sure you complete this exercise by 'taking care' of what you have produced, by finishing off this bit, keeping it safe and by separating yourself from it while acknowledging what hard work you are doing.

Part Three
The Separation: Karma From Dogma

Chapter Seven

Here's One I Made Earlier

"After you was born, I felt whole, perfect. Then Dad came in and I felt dirty again…later…I went a bit wild. After he found out I was pregnant he just put me down as a slapper. I didn't want to disappoint 'im again did I"
Kat Slater to Zöe Slater: Eastenders BBC1 2.10.01

Opinions differ about whether Kat of Eastenders really is a 'slapper' or not. Personally, I liked her even before we found out about her abuse, particularly once she stopped wearing bright orange make up. Perhaps this is because I felt I understood her anger and her behaviour. In this chapter, I want to look at some specific examples of how we 'make' realities as well as looking at why we might be making them in that way.

To avoid getting silly about this subject it is best right now to separate natural disasters from dramas we participate in. People who say that, on some level, those in the World Trade Centre willed what happened to them or those that died in the Hatfield rail crash chose it; those who say children who are abused or people born with learning difficulties or deformities, or people with cancer wanted this to happen on some level (and people really do say these things) are simply doing more damage to those already suffering. I had a client recently – let us call her Sue, who as we looked more closely at her experience, saw that had she made some different choices in her life, she would have suffered less. Then she used this as a reason to blame herself. This is typical of people who have been blamed and abused. They become over responsible in an attempt to retain a sense of control over their lives. The nasty behaviour visited upon her by those around her was down to them, not her, and given her youth and situation, she could have no more stopped them from behaving that way than the commuter on the train through Hatfield could have stopped it from crashing. True –he might never travel on a train again, and will doubtless always be on the look out for signs of difficulty if he does, but ultimately, if he trusts, he's vulnerable and if he doesn't trust, he's still vulnerable. As the ever wise and compassionate

Dumbledore points out:

> *"The consequences of our actions are always so complicated, so diverse, that predicting the future is a very difficult business indeed."*
> *Harry Potter and the Prisoner of Azkaban:J.K.Rowling*

And he's a magician!

Sue is also still vulnerable, but she's a bit more in touch with her feelings now about what she will put up with and what she won't, and this will have an impact on her future. This is the whole point of experience and choice. As children, we have a simple, unsophisticated statement: 'I wonder what will happen if I do that?' By the time we are adults we have a much better idea about the basics but no clearer knowledge of natural disasters.

The process of therapy involves looking at how we got to where we are. As we deconstruct or take apart our past, we get to see how it was put together. Like a car engine with a blown head gasket or a washing machine with a blocked filter, we repair the bits which keep sabotaging us and making our 'life force' oil leak away and we clean out the fluff and the rubbish and soften the water we use to keep the system going. But first we need to know where the block or the leak or the diversion is.

Erroneous Assumptions

In chapter four we looked closely at assumptions and how they affect our lives. We looked at how to access our own and picked out some that are quite common. They were:

- Nobody cares about me
- No one is listening to me
- People with more power than me will abuse that power
- I don't have any power
- I am a very bad person
- I am worthless
- I don't even exist
- If I try harder perhaps I will become worth something.

For *Eastender* Kat Slater, abused by her Uncle Harry, it's not difficult to see how these beliefs became established. They were a way of making sense of her experience. Often our beliefs are established early in life and become attached to situations far removed from the original trauma. The more entrenched they become, the more difficult to see or to change. My own experience as co-ordinator of an agency providing counselling for people who had been sexually abused provided a salutory lesson in this for me. As both counselling supervisor and fundraising co-ordinator, I was responsible for the volunteer counsellors. Had I realised at the time that the level of training and experience was very basic, I would have been far more supportive and a lot less challenging. I had assumed they were (although volunteers) professionals who could meet me, in all of my complex and high powered job description, warts and all, on a equal basis. However, what some of them were after in their indirect move into the mushrooming world of counselling, was a mother who would meet *their* needs. They were not aware of this and, unfortunately, neither was I. Because this was not recognised, what some volunteers experienced was a 'not being taken care of'. Believing themselves in the right, (because being a good counsellor means being nice to everyone doesn't it?) they aimed their feelings of envy and hatred and oppression at me. In this chapter I shall track a little of what made it possible for them to do that as an example of just how much of our reality we make up and how, but first perhaps it is important to show how I came to understand that.

Years ago, as a lowly worker in a large social work organisation I fought very hard when I felt unheard and accused one manager in particular, of being oppressive and of using public school bully-boy tactics. At one point, I even thought that the clients and the untrained and unskilled workers should make all the decisions. While I still believe that these are the people who know best about the kind of service that should be delivered, it would have done no one any favours to have them hold all the power. In the same way, to allow my own hatred and desire for revenge from long ago loose on my manager would have been disastrous yet this was what was fuelling my problem with authority and my fear of power being abused. If people have been oppressed or abused enough it is not difficult for them to justify and find excuses for ruthless behaviour which is actually fuelled by emotions and assumptions which have very little to do with the current situation.

In my case, my manager refused to really listen or to engage with my views because he was a man unused to a young tactless woman with radical ideas. Since my father never listened to anything and always shot my perspective down in flames, crushing and dismissing me as irrelevant and stupid, my manager looked a lot like him and so set off identical feelings of powerlessness and rage. However, he was not my father and he was not actually trying to destroy me. He was simply not very good at that aspect of his job.

A few years later I got a management job myself, and all the stuff I had accused him of, I was accused of. The irony was, that even though I became aware of what was going on and took great risks myself to provide what was needed, they had the power to crucify me if they wanted to, and they did. Had they been able to separate *their* anger from past oppression (This was an agency dealing with sexual abuse every day so there was a huge amount of distress and disturbance around) they would have examined the facts that didn't match their beliefs as well as the ones that they *thought* did. The pain of separating these things out, as we shall see here, is profound.

For Kat and Zoe, the *Eastenders* we thought were sisters but turned out to be mother and daughter, there is a powerful scene where Kat tries to tell Zoe what happened and Zoe struggles to make sense of it within the framework of her limited experience:

> *Zöe: "The reason those things were happenin' to you is because you did what you always do and let blokes use you."*
> *Kat: "It wasn't like that."*
> *Zöe: "What was it then? Were you raped?"*
> *Kat: "He was older than me"*
> *Zöe: "That's not what I asked you."*
> *Kat: "You don't understand."*
> *Zöe: "Either you were raped or you weren't"*
> *Kat: "Did he put his hand over me mouth and hold me down? No. Did I tell him to stop and scream for help? No. But I was raped. I was thirteen years old. What do you think?"*

Zöe: "If you didn't want it to 'appen why didn't you stop 'im?"
Kat: "Don't you think I've lived with that all my life since? Don't you think I've asked myself the same question every night and there's only one answer. I must be the slapper everyone says I am."

<u>*Eastenders BBC1 2.10.01*</u>

Zöe's assumption is that if her mum slept with someone at the age of thirteen, she *must* be a tart and a whore. That was the standard belief about women for years, though thankfully stringent tightening of the law to prosecute older men who have sex with girls indicates a waking up as to where responsibility lies. Not in the East End though, especially at the start of the 1980's. What we see going on in this conversation is how Kat came to the same conclusion about herself, *despite* her experience. Even though she struggles to make it make sense, a part of her says:

"I must be the slapper everyone says I am."

She speaks like a whore, she dresses like a whore and she behaves like a whore. By taking on this assumption she becomes that 'slapper', even though part of her knows it's not true and that she is more than that. Abused at the age of 13 by her Uncle, Kat has spent the rest of her life so enmeshed in society's beliefs and in particular, the other most powerful agent of influence apart from the mother (yes, it's Dad's again), that she is unable to stand up and put her experience; her own, real, traumatic, agonisingly powerful experience, consciously at the centre of her life.

Just as Zöe cannot grasp any possibility other than to blame Kat, so Kat had little chance to build on and be supported in the truth of what happened. With one sweeping wrong assumption, her Dad destroyed what tiny part of the truth she might have been able to share:

Kat: "The worst part of it all was the look on Dad's face when Mum told 'im. I thought 'ed hit the roof, but he didn't – he just looked at me with contempt...

he said he was disappointed in me. Thirteen years old and I hadn't lived up to his expectations. I'll never forget that look. I just wanted to be dead. I didn't want to be here anymore. Nothing he said or did after that could take the feeling away."
<u>*Eastenders BBC1 2.10.01*</u>

Her Dad's approval was so important that Kat didn't even want to exist anymore. Her sense of her own worth was that fragile, even at thirteen. In its absence, her reality was created by the way others saw and classified her behaviour:

Kat: "Good old Kat. Quite a laugh in't she? Always up for it an' all. No one does it like me, eh Zo?"
Zöe: "You said it."
Kat: "Everyone else says it so it must be true."
<u>*Eastenders BBC1 2.10.01*</u>

Finding Evidence

So, if everyone says something must it be true? Kat only has to look at herself now to strengthen the lie and we'll see more about how she does this in the next chapter. In life, as in science, we form our hypothesis, or belief and then look for evidence to back it up.

For myself, with my 'oppressive' manager, the evidence was in his bombastic and forcing behaviour. I once had a row with him in his office where I was verbally assaulting him for referring to someone who had recently been in care as a 'wide boy'. For me this was evidence that he related to people in a stereotyped way and should not be in a position where decisions had to be made. He raised his voice to get a word in edgeways and I was so appalled by what I saw as these 'underhand, manipulative and typically paternalistic' tactics to shut me up that I walked out of his office and he followed me down the corridor shouting, "COME BACK HERE". Well– for me that proved beyond a doubt that he was a megalomaniac who would stop at nothing to retain his absolute control of me, of the kids, of the project, no matter what it said in the so-called enlightened policies. Here was the proof! By this time I saw this man as a personification of evil rather than an old fashioned man ill equipped to cope with my approach and who had fallen back on what he knew when he felt threatened. In an after-the- horse-has-bolted kind of way, that man did do more training after I had left and I was told that he became a lot easier to work with.

In the agency working to provide counselling for a minority client group, I was a whirling dervish of hard work. All aspects of my job were linked. Hundreds of thousands of pounds were needed from the National Lottery and to get that takes outlining *exactly* what an agency

...has achieved
...hopes to achieve.
...will cost to the last penny.
...will provide for all of the clients.
...will include for non-clients in the community.
...will provide in terms of training for other agencies.
...will provide in terms of training for volunteers.
...will provide in terms of policies for every occurrence
And exactly *how*

To keep money coming in at all, the organisation needed to be developing new aspects and including more members of the community and constantly re-evaluating and rewriting its progress and its policies. Despite only working four days a week, I made huge inroads into all these tasks. Since the offices were painted a rather institutional magnolia, I decided they should be painted. I set up a box for suggestions of any sort in the office and sent round a survey asking for views and input on colours to all the volunteers and the paid workers. I received a response from around 10% and only one person was against my suggestion of the colour. As a very 'get on with it' person, I painted the agency corridor mauve, reflecting that if the worst came to the worst and it was vetoed, well, I could always paint it back again. The other paid workers chose colours for the counselling rooms and it was done. This painting later became the evidence for my absolute disregard of some members of staff. For a few, it made me 'look' like a dictator and while I acknowledge that inexperience, the very broad demands of my job and my own ability to focus on one thing at a time contributed to this 'feeling' or 'perception' it is simply not true that I 'was' a dictator. Clearly a few people felt so threatened that they needed to paint me as the source of all their difficult feelings.

Although the demands of my job description meant that direct work with the volunteers was a tiny part of my job description, this was not something they realised. The issues of powerlessness, a sympathy with that feeling and ability to identify with those who had been 'walked all over' were what had brought them to work at the agency. These few volunteers saw anything that was not primarily aimed at alleviating feelings of powerlessness as part of the problem – but I was not there solely to alleviate 'their' feelings of powerlessness, but to alleviate the clients and to make sure *they* were alleviating the clients. It is understandable how they came to *expect* me to be a source of endless loving warmth and sympathy but that was not my job unless I was actually supervising (which has great similarities with counselling). For these few then, while I was not providing support but doing something else, I became the enemy.* As a person trying and able to achieve a high turnover in organising, writing and grasping overviews (the tasks I was employed to do) I can be quite short with people out of the counselling/supervising context. This was enough within this underfunded and relatively inexperienced agency to cause real fears about what was *really* going on.

Just a few people felt dismissed, abused, oppressed. As manager and supervisor I provided further forums for opinions and feelings to be heard. I apologised for misunderstandings. I listened while I was accused of everything from bullying to malicious abuse. Even my painting was deemed to be full of violent power-mad messages done for the sole purpose of asserting my superiority. I was way out of my depth but I reflected, trying to learn and move forward. As I reassessed what was going on and realised I had miscalculated the level of reassurance that some volunteers needed it was already too late. Someone passed their version of events to the management committee and I was suspended for *gross misconduct:* that's how angry they were. This charge was dropped almost immediately since it was so obviously fuelled by an inappropriate level of outrage and the solid evidence to back it up just

*This is exactly why BACP guidelines stress that management and supervision remain separate. For many organisations this is a problem because of extra costs.

wasn't there. However, I remained suspended and this meant that I was forced to suddenly stop providing counselling for the clients I was seeing. In forming and keeping a commitment to me, these clients were in the process of coming to terms with long held feelings of betrayal and abandonment. Tragically, the sudden actions of the agency meant that the deeply held belief (that no one could be trusted) resulted not in being replaced with trust and healing but in being reinforced by the very agency supposed to offer an alternative experience.

I was put in the position of the utterly powerless and guess who got to feel powerful? It was not just the volunteers but the management committee, run by untrained users of mental health services. These people really had some axes, rather than skeletons in their closets and with good cause, considering how *they* had suffered in the past. They knew that people could abuse power and, since this time they could assert power, they found a target and made *me* the scapegoat. Their conviction that I was oppressive and bullying was so strong that the evidence of whether I was working to empower people (and there was plenty of this) was simply disregarded. I could produce suggestion boxes, consultation sheets and show that I had implemented every suggestion made, but no matter how I demonstrated the hard truth of the way in which the organisation had not been meeting the ethical conditions they paid lip-service to, no matter that I explained the policies that had to be implemented if the situation was to be rectified they simply refused to see or hear or trust me. I got a real taste of what it's like to not be heard, or seen; what it's like to be on the receiving end of denial and abuse and I don't think this was a coincidence when you consider the nature of the difficulties we were all working with. That is what a scapegoat does: takes all the bad stuff.

Ultimately, what was simply a passion for my work, pressure of time, and a determination to make the agency good enough for the serious nature of the work; what was an admittedly blind miscalculation on my part about the abilities of the volunteers was transformed into an unshakeable belief that I was a monster.

We hold such power over creating our own reality that out of a situation where my actual hard work had renegotiated funding and created the conditions to continue the agency, many people who had never even met me demanded my resignation on the grounds that I was seriously damaging it! And they had the power to force it, evidence or not.

Dreaming Our Dreams Away

Shamans are well known for their dreaming. When we hear about people dreaming the future we think of premonitions and ecstatic visions but dreaming our future *into being* is something that we do all the time. If we are unconscious and filled with fears, believing "*all men are bastards*" then what we find is men who are bastards. We filter out the men who aren't because we're not looking for them and we are often drawn to the ones who will prove we are right.

Mostly our primary experience with men is with our fathers. Rarely loving, my father was not nurturing or indulgent in any way and he provided very little approval. He found children a disturbance to his life and when that got on top of him he was not only critical and demanding but also sometimes contemptuous and full of verbal violence. Perhaps there was a remnant of this that clung to me in my managerial role and so enabled such simplistic projecting within the agency. Happily for my own sanity I was not alone as I was neither the first or last co-ordinator there to suffer and leave in adverse circumstances.

Despite having met many men who could not be more different from my father in their behaviour, their background and their beliefs, I have so far managed to turn each and every one of them into him to the extent that, if they weren't driven to exasperation by me at the beginning of the relationship, they were by the end.

What you know becomes what you expect, becomes what you get unless you can really see what you're doing and change it.

My husband has a dreadful memory. He gets tired easily and he doesn't seem to listen. For me this translates easily into 'he doesn't care' which could easily become 'he is with me for money/sex/because he can't find anyone else'. If I have to repeat myself to him, I take it as evidence that he has either failed to cover his tracks or that he is dismissing what I say as not being important. I am so concerned that he is out to abuse and take advantage of me in some way that I cannot see the simple reality that is him.

My previous boyfriend was so smooth that his ability to tell me what I wanted to hear dripped off him like honey. He told me he would love me forever and never leave me. I believed him. Since he had already left one wife and two kids (not for me) this was not very logical. Fairly early on in our relationship, he gambled the rent away. I began not to trust him. Always on the look out for being duped I became rigorously literal and logical. He once said that he would replace a bowl he smashed and that he would buy a bathroom stool. I still remember that and it was sixteen years ago!

I insisted that we instigate a cleaning system in the house to make it fair and to make me feel noticed. I kept making little bargains with myself: "If he won't do this or that, it will prove he doesn't care". He refused. He felt controlled. I felt unloved. I was determined to prove that he didn't love me. For me, my feelings of not being loved showed that this *must* be true. I embarked on a lengthy assassination of his character to prove how he was not showing me the love he would be if what he had said was true. He longed for approval, in fact, making women smile at him was more important to him than anything, so my strategy didn't work too well.

I longed to be taken seriously and felt this man was destroying me. I tried to make him feel the way I felt. I tried to get him to see that he was insensitive, fickle, shallow, inconsistent and persecuting. As you might imagine, that didn't go down too well. I'm sure it did make him feel how I felt but it left him unsupported in life with few options. Eventually I managed to make him become those things, just as he had pushed me into being the Wicked Witch of the West by not really valuing my needs and views. We really were perfectly matched. He didn't have the guts to say he wanted a wife who'd make his life easy and I didn't have the guts to say I wanted a man who would respect my requests. No – we had to find a much more complicated way to separate. The 'all-blame, long drawn out emotional suicide pact' as opposed to the 'no-blame quickie!'

He ran off with someone 20 years younger who appeared easy to be with. When he came round to discuss the practicalities he sat on the sofa opposite me and he said:

"I can get any woman I want."

And he meant it. The Machiavellian role I had cast him in was probably very convenient. It was the sort of thing people say when they have their back against the wall, when they feel so threatened that they have to believe they are invincible. We have huge resources available to us as to what we can become. We are not fixed but are all those things at different times: shallow, fickle, brave, funny, vicious. It is when we fix others in these roles that we restrict what we experience and eventually, create the reality we fear.

When I allowed myself to consciously dream of a man with gentle qualities and practical abilities; with hair down his back and confidence and warmth in his hands, he turned up. What's more, he fell in love with me. Too complicated and potentially painful for me to allow, it was short lived but this man appeared in my dreams at night as well as in the day, filling me with a security and a feeling of being loved that transformed my everyday experience into something very different from the suspicious, cynical terrain I usually occupied.

Perhaps these experiences just happen to provide an understanding. For me, the process of creating my own reality within my relationship is more an everyday experiment than an ecstatic vision. Like most of life, it has to be. It gives me a chance to examine what I'm saying or feeling or believing and put it against what I know. More often than not it shows me that it is more important to me to supposedly know what's going on than to allow myself to be open to the good stuff, the imperfections of my partner or the possibility of trusting what he says.

"*Maybe sometimes we don't do the right thing because the wrong thing looks more dangerous and we don't want to look scared, so we go and do the wrong thing just because it's dangerous. We're more concerned with not looking scared than with judging right."*
The Amber Spyglass: *Philip Pullman*

For me, fighting back and stating 'how it is' is a vital step away from letting my father's perspective walk all over me, but it isn't far enough.

As we have seen, our beliefs are often so strong that when we come across an experience that contradicts them, we ignore it. Kat in Eastenders was so entrenched in her role as 'slapper' that when the doctor challenged it by treating her well, she couldn't cope. Instead of changing herself and accepting him, she kept her hard shell and ditched him. Sometimes, when we really are struggling with a very tiny sense of ourselves it becomes more important to 'know where we are' even if we are deluded, than to look clearly at the evidence in a fair and balanced way. The monster within us prefers to feel strong and talk absolute rubbish than to be dissolved and experience weakness. We can dream our reality into being but more importantly, we can dream our dreams away without even noticing we're doing it.

Exercise

Permission: 5

The false selves that we have built up become increasingly entrenched because we keep doing things that prove to them that they're real. Everything becomes grist for the mill of our limited view of ourselves, but once we see this clearly, we have choice. The child we met at the end of the last chapter, is of course, us, and the imaginings an attempt to get a different perspective on ourselves. It is when life gets difficult and our stress levels rocket that these automatic messages, and often the behaviour, kicks in.

To complete the journey into yourself that began with you writing down your automatic messages you will need a pen, and your box full of messages. By now I hope that what was automatic has been replaced with a healthier automatic 'taking care' – making sure you are safe and won't be interrupted.

1. Think of the last time you got angry or upset. What happened? Did you spiral into a concerted attack on yourself and end up with the conclusion that your core messages were true?

2. Imagine this happening to the child you love. What would you say to it at the start, before the cycle took hold? Perhaps it would be: 'You're tired. Why don't you have a short break'; or ' Yes, it's really

difficult'; or, 'Well done for keeping your temper'; or, 'Considering the circumstances you're doing really well'; or even 'Yes, it's completely chaotic.' Take your time and consider what you really want to say to alleviate his or her suffering and to put the incident in perspective.
3. Write it down as a finishing of this process that began when you first started to get in touch with the automatic assumptions or messages. Take each message and find the heartfelt response inside you to this poor, battered child. Say it to yourself. It might sound corny but perhaps that's the you that sabotages your best interests by not allowing you to do what is *really* good for you.
4. As you improve in your ability to do this exercise, do it whenever you are having a particularly bad day. I guarantee you will short circuit the pain and reduce it by about two thirds.

Chapter Eight

The Vicious Cycle

"I don't want you to say it's wrong. I just want you to cuddle me like I wanted you to cuddle me then. Where were you dad?"
"I was there"
"No you weren't. I needed you so badly then."
Eastenders: BBC1 4.10.02

And so it goes on. What will Kat expect from men in the future? That when she really needs a man, he won't be there for her. We take the realities and the feelings from earlier in our lives and transfer them to a new set of circumstances. We find 'enough' in the new relationship to stir up the same feelings and set off the old beliefs. Finding ways to analyse what is happening is one thing, but how do we keep the demands of the monster inside from destroying anything good that comes along?

Anger and Guilt

Our anger and our guilt has a huge part to play not only in creating the reasoning that keeps us running around the hamster wheel of our delusions, but in keeping us there at all. If we can grasp this we can break the cycle and if we can break the cycle, we can step out of the game any time we notice what we're doing. I always thought therapy was about reliving experiences or confessing my bad side. I was in there for 5 years, twice a week for at least half of that time, yet the most valuable thing I learned, right at the end, after I had decided to stop going, was tiny. One tiny insight realised in one moment that created the only crack I really needed in the bricked in wall of my vicious circle. One that many mothers, in particular, can probably recognise.

It was a normal rushing working day. Take daughter to school, go to work, get shopping, pick daughter up, cook dinner, eat dinner, get daughter off to Brownies by 6pm. I was a single parent and sole breadwinner. I had told my daughter to get her handbook and her coat ready so that we could leave as soon as she had finished watching

Neighbours (I was doing whatever else it was that needed to be done). She ignored me so that when Neighbours finished she had to go and look for them. Stressed, pressurised and tired, I was not pleased so, not trusting myself to speak to her, I got in the car and had to beep the horn twice before she got in it. We drove the 2 miles in angry silence, she got out and slammed the door and as I pulled away I caught my inner voice saying:

> *"God you're a terrible mother. You're horrible, violent, aggressive and will mess her up totally and it will all be your fault. You should never have had children. You didn't deserve them. She'll be damaged for life because you're just like your father."*

And then I realised that this was the kind of self-attacking sweeping statement claptrap we had identified so clearly in the therapy room. It just wasn't true. The truth was that we'd had a battle and a bit of conflict, which we'd both survived. I'd asked her to do something perfectly reasonable and she hadn't done it so I'd got angry. So what? This did not make me a terrible person. It illustrated that I was a person who meant what I said and who had limits; that I was not a saint, a martyr or a person capable of providing endless unconditional love. That I was just an average human being. That I might even be 'a good enough mother'.

So the guilt at being 'bad' and the angry part which knows I'm not bad and that I have rights too, were doing a great job at sabotaging this otherwise well-balanced woman. The cycle works like this:

1. The feeling of guilt means I try too hard to be good and take all the blame for everything not being better.
2. I put up with being shoved around by other people's endless demands until I'm really angry.
3. The anger isn't just with what's there but is all the anger I ever felt from being pushed around in the first place.
4. I behave incredibly badly with the whole of this anger by losing my temper and hate myself more because after all, there's the evidence.

I was just absolutely appallingly vicious.
5. I'm obviously a terrible person so what right do I have to get pissed off with things? They are nothing in the face of my badness. I try harder to be good enough and hate myself more for being unable to become a saint.

It seems to be the way of things that people, however well meaning, will place unwelcome demands sometimes and be oblivious to the feelings, wishes or beliefs of others. Bearing this in mind, major anger and guilt problems need to be seen for what they are if we are to defend ourselves from the demands, conscious or otherwise, of others.

Damage

"Have you ever been scared Zo? Ever been so scared that you open your mouth to scream and nothing comes out? When you see something so horrible that you turn your head away and pretend it aint 'appenin'? Don't you tell me what I should've done. You wasn't there."
Eastenders BBC1 2.10.01

Trauma. How do we work with an experience of non-existence so serious that rage is the only thing that makes us feel alive? In Kat's early episodes of Eastenders we see rage fall out of her at every possible opportunity. Kat rarely does anything but tell people where to get off. While on the one hand she feels justified, how on earth is she supposed to stop doing that and what would happen if she did?

My guess is that she either exists in a loud and unavoidable way, or she becomes smothered by feelings that are very much the opposite: feelings of not existing. These feelings of not existing lead to panic which provide the energy to be violent and abusive herself. It seems as if the forcing or obliteration of Kat by her Uncle Harry has left her with these same tools: to force and obliterate. She is so concerned with being able to breathe again herself that she is oblivious to the impact her abusive behaviour has on those around her. Like a person drowning or suffocating, escape becomes everything. She *has* to avoid feeling this bad again, she *has* to avoid feeling as vulnerable as she did with Harry; as if she does not exist and has no power at all.

Why is this so traumatic? Why is abuse such a big deal? Abuse has become such a well-documented issue where once it was a taboo that to ask this question has almost become taboo in itself. In relation to Kat, we are discussing sexual abuse but emotional, sexual and physical abuse have more in common than you might think. The impact that they share is not to do with what happens to the body, in all these different manifestations, but to the mind.

Any violation of human rights has a price. Let's consider:

- The right to free thought and speech
- The right to keep one's body untouched by another
- The right to not be beaten

When these are violated as in a Chilean torture chamber at least we know who the enemy is. When it is someone bigger than us, who we have always thought was wiser than us and who we love, what is being abused most is our mind, our whole sense of what is real. We trust that we exist, but having just been treated as if we don't, a doubt creeps in. If we have been treated badly, where is the reflection, the compassionate image of our suffering that would give us a sense of ourselves, a sense of our rightness to suffer, a sense of acknowledgement?

"He kept sayin' 'it's going to be alright. Don't you love me?'
When he'd finished, he cried. The next day he just carried
on like nothing 'ad 'appened."
Eastenders BBC1 2.10.01

The way Kat delivers this speech, it's like she still can't believe he did it. This dominant feeling of disbelief is one I frequently meet in the speech and the responses of my clients. Is it so surprising that faced with the truth, with the unbearable horror of what it means to be betrayed and unprotected, we choose to assume that it must have been our own fault? This way we feel as if we have some power. At least we can pretend that we will never let it happen again, because after all, it was down to us wasn't it?

Betrayal

As children we generally trust the people we love more than we trust our own judgement. If they say we are bad girls or boys, worthy of hate or beating or using like a blow-up doll, a part of us believes them, this is why we struggle to be good. If we simply can't face the truth of what has happened we become split into many parts, the bad; the good; the part filled with rage; the part that doesn't exist.

"I catch 'im lookin' at me sometimes and I still get them feelin's."
Eastenders BBC1 2.10.01

My hunch is that it doesn't take much to set off 'them feelin's'. In fact, pure logic would indicate that they will be set off by just about any man who shows an interest. The deep longing and need of the abuse victim to be loved set next to the fear of not being loved constantly reactivates such feelings and can set off behaviour that elicits very negative responses from people. In particular, if we lose the care and protection we need at a very early age, we find it hard to accept that if someone loves us and makes us feel so good, how can they not know how we feel all the time? The same feelings of confusion and disappointment besiege us now as beseiged us then. Our rationale becomes that they *do* know, they *must* know, but they're doing it on purpose, abusing power ruthlessly. This is what the people we love always do to us isn't it? The slightest hint of distance, or gap in reassuring gestures brings on the nightmare of loving, giving and being rejected and, in some cases, psychologically destroyed. Suddenly, and automatically, we might set into motion desperate behaviour aimed at making the person (and therefore the fear) go away from us at all costs. Rarely do we realise, feeling so invaded and powerless, that *we* could just go away, because, in fact, we are adults and we are free.

What Kat needs is for the abyss opened up by Uncle Harry's abuse and her Dad's betrayal to be mended. She knows the rip is in the fabric of a coherent caring understanding.

"I just want you to cuddle me like I wanted you to cuddle me then. Where were you dad? ... I needed you so badly then. All I wanted was for you to sit me on your lap and say it di'nt matter, and that you still loved me and everything was gonna be alright, but you di'nt, did'ya dad? You just looked at me with disgust. Disgust and contempt"
Eastenders BBC1 2.10.01

"Like a nightmare" is how one of my clients described events in their life, but the irony is that we often relive our difficult feelings in contexts that do not warrant this strength of feeling at all. When disgust and contempt become self-hatred and the fight to survive kicks in, we often exhibit behaviour that only makes sense in relation to what is going on *internally*. If the inner experience is of panic, of not existing or being seen we might do anything to rectify that. However, the way someone *outside* of me is *actually* behaving may be perfectly appropriate and benign. Sometimes if I get too close to a partner and feel I am disappearing, I would prefer to be shouted at than be left with the panic of non-existence so I provoke an almighty row. My husband is the most gentle man I have ever met (fortunately) so the possibility of him ever hitting me feels approximately nil. Nonetheless, sometimes my moodiness and desire to get away from him has been known to facilitate him into absolute rage and frustration. Once he loses his temper and shouts I am left feeling many things: terrified, triumphant, relieved, able to cry now and righteous, as if I have proved that my fears of him are justified. That he really is violent. In trying to prove that my perceptions are accurate I can act quite madly. In trying to prove not only that I exist but that I can be seen and have an impact, I have achieved my goal. In the long term of course, there is a limit to how much of this behaviour others are prepared to take, even if they love you to bits. Since the bottom line for survivors of abuse is that we are unlovable and that the person we are with is bound to leave in the end when they find out how disgusting and unacceptable we *really* are, this is a reality we collude in which is perhaps a life's work to change.

Doubtless within the counselling agency I referred to in the last chapter the fear was that authority would steamroller, be invasive and not value people. While the experience of being helpless in relation to that greater power might be 'just a feeling' – as it was for those volunteers who projected their fears onto me, the belief can be so strong and the feelings of shame and failure so extreme that they can't be borne. These internal reference points that can be activated by someone being a bit busy or not listening or even too happy can be a lot stronger than the evidence to the contrary. One beautiful spring evening I sat in golden warm sunlight looking at an abundance of flowering daffodils and they didn't touch me. The internal reference points telling me that there was nothing but despair; that it was hopeless, were stronger than the beauty outside. Blaming and misperceptions are bound to happen for people who have survived serious abuse unless their feeling and experience is explored and accepted. Only then can it begin to transmute into wisdom. For wisdom appears to be nothing other than digested pain.

Persecution

In *Nineteen Eighty Four* by George Orwell the real tyrant is not so much the political dictatorship but the language, '*Newspeak*' that the citizens are taught from birth. When people try to think freely they are unable to find an idea that will liberate them. In their heads they only know that trying to think of a new way of living is wrong, and this perception (that trying to think is wrong) becomes a brick wall in their minds:

> *"In Newspeak it was seldom possible to follow a heretical thought further than the perception that it was heretical; beyond that point the necessary words were non-existent."*
> *George Orwell: Nineteen Eighty Four*

In many ways *Newspeak* is a metaphor for how we think all the time. Our understanding is restricted to what we *think* we already know. The 'knowledge' that witches had moles and lay curses on people meant that if someone died and her neighbour was in any way unusual, it was a good idea to look her over for a mole. If one was found, this was evidence that she should be hanged as a witch. That's logic. It gets applied to what we think we know in order to find an answer often to questions that are too complex for us to *know* the full answer.

So what should we do if speaking freely is perceived as oppressive? Take, for instance, a general climate of political correctness within which we can easily believe that loud or opinionated people are, *in fact,* oppressive. There may be nothing illogical in our thinking, the problem may be that we are working from limited data. Data that says – *if I feel oppressed, it is because people are oppressive,* rather than seeing that the person in that role or mode stirs up difficult feelings. The next step (from this false conclusion) is to follow a policy of controlling, excluding people and/or dictating what their behaviour means. Sound a bit Gestapo-like?

Let's be clear here. The right to be safe to speak your mind if you *feel* oppressed needs to be available. Really available. Inscribed in law if necessary, with mechanisms in place to ensure it. This will safeguard freedom, and that is our concern within a democracy. Surely though it would be a mistake to adopt the more simplistic notion that having an opinion and the ability to argue *is* oppressive. If we build rules and sack people on that basis we create the deeply ironic opportunity for the victims of past *real* oppression or abuse to become persecutors as a result of generating their own dogma from their own experience.

People often think that what they are doing is fair and just and even heroic, but if it involves blaming, labelling and stereotyping of someone else, reflect fully before acting. The road to hell is paved with attempts at liberation. In the encounter groups of the 1970's, the most sensitive, open and the emotionally courageous person generally took the blame for the shortcomings of everyone else. When Anne Robinson of BBC2's 'The Weakest Link says *"You are the weakest link – goodbye"* she means it. It is a very accurate reflection of how society functions, particularly the commercial world. I would describe my clients mostly as the weakest links in a dog eat dog world. So is this a bad thing? They may appear to be inadequate people with 'problems' but they are often the most sensitive, aiming to mop up the world's troubles by naming taboos and injustice. Fundamentally trusting they often place potentially dangerous information in the hands of those people who will use it for their own ends because they think most people have the goodwill and openness that they have. Often they are stuck with a habit of absorbing the needs and feelings of those around them that they then try to 'solve' or 'make better'. They then become vulnerable and can easily 'absorb' responsibility, which many people are often only too happy to give them. This makes it possible for one person to 'carry' all the bad, and all the blame. Families do this a great deal; they ridicule each other and pass their own pain round like pass the parcel. The weakest link is the naïve one who isn't selfish enough or ego driven enough to protect their interests by withholding information, lying or omitting the truth; the one willing to put their head above the parapet and get shot down; the one who has always been so convinced that everything wrong around them is their own responsibility that they will allow themselves to take all the blame and end up crucified.

Control

Kat's attempt to maintain some semblance of order and control means keeping the world, intimacy and pain at bay. For seventeen years she has kept the family secret that would blow apart the illusions that the Slater family hold about one another. Her carrying of all the blame in the family, acceptance of responsibility for all its weakness shows a sacrifice that other family members were willing to benefit from for the sake of a comfortable life. Paradoxically, sometimes the weakest link can be the strongest link but only when they are fully free to choose that role. By not telling Charlie (her father) about Harry (his brother), Kat holds all the damage in the family so that no one else can be hurt and they're not keen to change that situation. Even when Kat tells Charlie that she had told her dead mother, Viv, about the abuse and Viv had brushed it under the carpet, Charlie just can't take it on:

> *Charlie: "She would've told me…she wouldn't have kept somethin' like that to 'erself, not from me"*
> *Kat: "Alright. I lied. I quite liked it."*
> *Charlie: "Shut your filthy mouth."…*
> *Kat: "That's 'ow you see me isn't it?"*
> *Charlie: "No."*
> *Kat: "Yes it is. Ever since you've found out I was pregnant. I*

could see it in your eyes. Oh you've tried to 'ide it. Pretended it wasn't 'appenin' but it's never gone away."
Charlie: "Not true."
Kat; "Yes it is."
Charlie: "Well yes. I was disappointed. You were pregnant at 13."
Kat: "It wasn't my fault."
Charlie: "I know that now and I would've known then if you 'ad'nt lied."
Kat: "I told mum."
Charlie: "So you say, but she wouldn't 'ave kept that to herself. She worshipped you girls. She must 'ave misunderstood what you were sayin'...There is no way she would've let anything like this go by. Not my Viv."

<u>Eastenders BBC1 4.10.01</u>

A typical case of someone not being who we thought they were. It happens all the time and here we see Charlie going round and round in circles, trying to make the pieces of the muddle fit together so that his belief about who Viv was (saintly probably) can stay intact. In order for that to happen, Kat has to remain at fault, carrying the blame as if it was pass the parcel and the music stopped 17 years ago. To maintain this fiction, to keep everyone static in their place, Kat has to be the kind of person with a 'filthy mouth' who lies, the very opposite of the description which cannot fail to blow this cosy myth right out of the water:

Kat: "But what about me?...I was thirteen years old and I 'ad my Uncle coming into my room at night and puttin' 'is 'and under my covers. Can we talk about that first?...I turned the other way, 'cos I couldn't believe what was 'appenin to me and I seen the moon and I knew that you lied to me."

<u>Eastenders BBC1 4.10.01</u>

Charlie just doesn't want to hear this and certainly doesn't want to connect with the pain Kat is experiencing. When Kat was a little girl he had told her that whenever she could see the moon, she would be safe, so she feels doubly betrayed by him. Connecting with his real love for her, his acceptance of her, would heal this wound and she knows it:

Kat: "It's not too late Dad. You could give me
a cuddle now. I could be your little girl again.
I'd still believe in 'im." (The man in the moon.)
Eastenders BBC1 4.10.01

The irony is that Kat *can't* go back. To believe again in the man in the moon would be a betrayal of her own experience. Cuddling Kat like a little girl would mean that Charlie would have had to recognise his collusion in maintaining the lie. He would have to apologise to take some of the lonely pain away from her. We want our world in manageable bites, where we seem to have some control over the chaotic, messy, ever changing confusion of its deeper nature. Why am I writing this book? Because I like to know what's going on! I like to be definitive, to know what I'm dealing with. Charlie can deal with this new knowledge about Harry, but he can't cope with his *feelings*. They are confusing and threaten to overwhelm him, so, in relation to Kat, he keeps them at arms length. He simply can't afford to see her as the vulnerable child she was. She is an outsider, abandoned again.

The less we show of our emotion and the less open we are about ourselves, the more power we have over others if we wish to use it. If others have information about us, they have power they can use against us. Be aware of who you tell and what you tell them. Part of the charges against me at the agency was that I 'disclosed' information about my own abuse 'inappropriately'. If abuse is part of your background it can colour your life and constellate your world and your family in ways you couldn't dream of, distorting reality until it is unrecognisable and creating an outlook fuelled by self-blame and over-responsibility. We are not responsible for what *has* happened to us if someone is abusive or violent, we are responsible for what we do about it *now* as adults, or for what we do not do.

Our responsibility is perhaps better expressed as our 'respond-ability', our ability to respond in an immediate way, to name our experience and to assert our rights is a healthy level of control. We may not be able to anticipate the outcome but at least we are back on our heroic path.

By contrast, unhealthy control is often about controlling others. We pigeonhole them with assumptions that keep a pain that we believe is intolerable, at bay. The same goes for our controlling assumptions about who we are. For Kat, the trauma of being wiped out meant she adopted Charlie's belief system to try and establish some substance. She betrayed and hated herself but no doubt it seemed easier than telling the truth and risking the entire edifice of the family. No doubt she thought that she was very bad much of the time, but lies can appear easier than the truth. Imagine a choice between the belief that you are bad, where there is always the possibility of becoming good and the experience of screaming endlessly for food or attention or just contact and no one comes. Eventually we will stop screaming and the experience that there is no hope will overwhelm us. As a vulnerable baby or child, this is too much to bear,* though as a robust adult choosing to share our experience fully with a therapist who will not leave, there is nothing to fear except fear.

Pain

We have already explored how trying to avoid pain actually prevents it from going away. As we experience ourselves allowing our pain, it gets easier and feels safer to do so but getting to the point where we can do that, where we can see ourselves clearly, compassionately and without judgement can take time. It is often a chicken and egg situation. In *The Snow Queen*, one of Hans Christian Andersen's fairy tales, Kay and Gerda have played and grown up together in their shared tenement building. One day a stranger sweeps through their village. It is the beautiful but cold and heartless Snow Queen who has come down from the north. Unbeknown to either of the children, up in the heavens some demons have accidentally smashed a mirror and the slivers of glass have fallen to earth. One of them enters Kay's body through his eye. As a consequence he can no longer see clearly.

When we talk about seeing clearly within psychology, we usually

* See Klein in the appendix

mean inner seeing, or the ability to see the truth as we experience it. In *The Snow Queen* this inability to see clearly is compounded because it is a *mirror* which is smashed. Specifically, Kay can no longer see himself clearly, only a distortion. He loses touch with what is important, and zombie like, falls under the powerful, malicious influence of the Snow Queen, who wants him for herself. A fragment of glass works its way right to his heart, the symbol of being able to give and receive love, and there it freezes him until he no longer knows who Gerda is or that his real home exists.

In the end, Gerda goes to find Kay as part of her own heroic quest, all the way to the North Pole. Helped and hindered by a variety of animals and plants along the way, she eventually learns more about herself and knowing herself more, becomes more comfortable with herself. When she finds Kay, she bursts into tears and weeps all over him, connecting with her feelings of hopelessness and despair. This washes the mirror fragment from Kay's eye and as he also begins to cry, so the pain and the obfuscation that was causing the pain (the broken mirror) becomes washed out of his heart by the tears. He can see again and they make their escape home.

This hidden connector is what counsellors are generally searching for. When a client bursts into tears and says things like "exactly" or "you've put you're finger right on it" or just cries and tells you more, I know I have helped us move forward. We need to remember and we need to see clearly and sometimes it takes a shock to help us get over our barriers and judgements. When Kat really sees the bad in Harry and names how he has blamed *her* for her abuse at his hands, she breaks the spell. She becomes free to change her patterns of behaviour, though it will take a lot more work on her part, and a lot of support. Although her feelings may be loosened up and transformed, this will not happen overnight and it will not be easy to avoid old habits.

> *Harry: "You really hate me doncha?"*
> *Kat: "Hate you? No. I just think you're a fat old perve."*
> *Harry: "You loved me once."*
> *Kat: "I loved Father Christmas an' all but he didn't touch me up"*

Harry: "It wasn't just me. You were always dancin' around in that little mini skirt."
Kat: "That was me school uniform."
Harry: "And the make-up."
Kat: "I used to walk round in me mum's high heels when I was three. Up for it then was I? You see. You can't hurt me anymore Harry"...
Harry: "You bitch."
Kat: "What? You mean I ain't your special little girl anymore?"
<u>*Eastenders BBC1 4.10.01*</u>

Now that the real perpetrator has been named and the anger has been used to make it stick beyond the shadow of a doubt, Kat is free if exhausted. Harry's games and lies, Harry's *reality,* won't wash anymore. Perhaps this sorry chapter would have been over if Charlie could have cuddled Kat to really give her the experience of being clean and free of doubt or blame, but she is still the weakest link, absorbing the bits Charlie can't take on. So the alien still lives within her. Unable to cope any longer, Kat overdoses and it takes this for Charlie to truly become connected to what matters:

Charlie: "My little princess. You used to wake up every morning singing...you see, you're so much like your mum and that's why I think I took it so badly...you didn't let me down, it was me who let you down and I'm not clever enough to find the words to say 'ow sorry I am about what 'appened, but I'll tell you this. I'll never let you down again. Never."
<u>*Eastenders BBC1 5.10.01*</u>

Finally, the Slater's have got to the bottom of things and what they've found there are tears, tears that will heal the pain, now that the whole sorry mess is out in the open. The more we cry, the easier crying becomes. The more we can articulate our feelings and our assumptions without judgement, the more healthily sorry for ourselves we can be, the less brutal and judgmental and the quicker we can cry and let it all go. For the time being. In fact, crying alleviates anger, irritation and disconnection and is a lot better than an aspirin. If your way in is to connect with some unpleasant assumption (here's some I made earlier)

- I can't do it.
- People will hate me.
- I'm not good enough

Then face them; feel them; see them for the codswallop they are and have a good cry. You *will* stop. I promise.

Exercise

Proof: All In The Mind

This is a simple and pain free exercise that can be done anywhere. It demonstrates the power of our mind to create our reality. You will need another person to help you.

Once you have got hold of them (this can be anywhere and with any genial acquaintance as it does not involve any delving), pass over this book and get them to read the following. (Do not read it yourself or it won't work!)

1: Get the person you are helping to hold their arm out to the side so that it is horizontal. Now ask them to think of a time in their life where they felt terrible. As they do this, push down on their arm.

2: Ask them to put their arm out horizontally again but this time to think of a really great time when they were really happy. As they do this, push down on their arm.

In my experience, number one went down easily whereas pushing down number 2 was much harder, which goes to show that nothing in 'reality' changes, yet undoubtedly what you think makes you stronger or weaker.

Part Four
The Integration: Magic Now

Chapter Nine

Create Your Own Reality

"Magic is the ability to change consciousness at will"
The Spiral Dance: Starhawk

We have gone to great lengths in this book to deconstruct what we do and how we create our reality. We are no longer children reacting to perceived slights. We have put away childish things, including the fairy tale dream that we can snap our fingers and effortlessly conjure what we need. Yet we have learned that with an immense amount of discipline and hard work we can move our minds from one reality to another. As with all deep magic, there is immense danger in this practice. Unless we do the whole journey and experience some of the real depths, we can fall off our path, lose our roots and whirl away like the girl in *The Red Shoes*, losing our connection to the things that formed us. Some are good, some terrible but we need them all if truth is to avoid becoming a dogmatic one dimensional image of desire, rather than all that we are.

"You can do many things. You can refuse the power you feel in you now. You can take it for yourself. You can fight it. You can throw it away. But whatever you choose to do, know that the power doesn't begin with you, or end with you"
Firesong: William Nicholson

We might be able to shift our consciousness and be occupied with and by what is basically healthy, but we live within a wider context with a portal at either end. These portals go by the name of birth and death. Over these we have no control. Forget this and you forget everything that makes life rich, meaningful and ultimately mysterious.

The Way of Things

"Every single being who knows of your dilemma wishes things could be otherwise: but there are fates that even the most powerful have to submit to."
The Amber Spyglass: Philip Pullman

Just imagine if there were no limits. If, when we lost our tempers or got out of control we really could change the course of the world and destroy everything familiar, everything we know. It is really quite reassuring that we can't manipulate facts, that despite the growing post-modernist belief that there is no truth about the nature of things, there clearly is. William Nicholson's trilogy *The Wind On Fire* is not, at the time of writing, as well known as Philip Pullman's trilogy or C.S.Lewis's tales of Narnia, but Nicholson's insights into the nature of love and separation, of self and other, of grief and living lead to the same vision. His powerful hero Bowman is angry when he finds his expectations thwarted and hurls all of his hatred at his sister Kestrel, blaming her as he searches to destroy her.

'He found himself tumbling into her mind, through
curtains of longing and fear and ferocity and love, through
the part of her that loved him, into a far place he had
never known before, where there was a vast stillness.
Here his attack ended. There was nothing to strike, no enemy
to subdue, no prisoners to take. Instead it was his own furious
power-driven mind that was subdued and taken prisoner.'
Firesong: *William Nicholson*

He has found nothing but himself and the feelings generated by his own envy and rage. These feelings were so powerful that he believed it was she who was in the wrong, but the fates have spoken and chosen her to do a task which he had thought was his own. The truth is we can only change that which we have created: the constructs of our own mind. When contemplating the great losses to come, Pinto, Bowman and Kestrel's younger sister is told,

"People weep for a while, and then they stop."

But this is more than she can imagine and she rails against it:

'If they never come back, thought Pinto,
I'll weep and I won't stop. I'll cry myself to death.'
Firesong: *William Nicholson*

But ultimately, when the great changes come she is told

"You have to look after Ma and Pa. You have to be strong.
Don't cry"...so Pinto tried her best to stop crying.
Firesong: William Nicholson

Of course, she could cry forever if she *chose*, but it would take a major act of defiance. She doesn't want the losses, but she can't change them, only her own attitude. Death is absolute. It is unchangeable. Yet if it wasn't there, if we had no boundaries or fixed things, how would we define ourselves, our likes and dislikes, our sense of who we are? How would we grow into who we want to be? It is the way of things that we can't destroy everything we don't like but we can look at our beliefs around them. Often we believe that we can bargain with God. We believe that God will reward us for goodness and prevent us from experiencing tragedy. This is an illusion that prevents us from facing just how little control we actually have over things. The truth is that the systems we are surrounded by are not personal but particularly impersonal and bigger than we are, even the smaller ones that generate bills and try to sell me things I don't want over the telephone. To add a humorous note to a serious subject it occurred to me one day that if I was *really* being as persecuted as I sometimes feel, I would get final demands for gas and electricity stating:

Dear Ms Jackman,
We are very disappointed that you haven't paid for your gas.
We really go out on a limb to send out these bills. It's not fair and if you don't pay now we'll cut it off. It costs a fortune to send out these final demands so what are you going to do about it?
It's time you grew up and took responsibility, you're pathetic.

To follow this up they would, of course, cut off my gas in midwinter and laugh when I asked them to put it back on!

The way to get rid of illusions is to not assume anything, yet one of the things about being human is that we learn through a series of assumptions.* Like scientists we posit hypotheses. It is mainly guess-

* See Piaget in the appendix

work that gets us through, but if we want to expand our views into what is possible we have to recognise guess-work for what it is and check out where our reality fits with another person's. My husband not phoning or remembering I have an interview doesn't mean he doesn't care although that doesn't stop it from hurting. I may stop caring myself, in order to become protected but then I would just be deadening myself. He may or may not care but this won't stop him from being his forgetful self. Only he can make steps towards changing that and he's far more likely to do that if I tell him how it affects me and stop pushing him around and blaming him.

Our assumptions contain an enormous amount of power. When Bowman comes to liberate a city from Albard the tyrant, Albard responds:

> *"What makes you think they want to be free? You think I compel their obedience?...I am what they have made me...you've come not to free them, but to free me."*
> Slaves Of The Mastery: *William Nicholson*

Escaping from the confines of belief systems takes more than looking at the way we feel oppressed. We are all tangled up in the assumptions of others as well as ourselves, whether they are weak or strong, rich or poor. In taking the roles in which we are cast as fixed, we unconsciously collude in an illusion while remaining unaware of the magic that holds it steady; the tacit agreement which enables it to continue with ease. There is always far more to things than is apparent because the composite parts are either growing or blocked from growing and our way of looking at them and decisions about them are created by ourselves and so are potentially endlessly changeable. I know people who have characteristics that make me fume. They are never on time, don't do what they say they will do and absolutely *appear* not to care, yet they are consistently gentle with children and redeem themselves by the sense they talk or the overt depth of their souls. What is more, they refuse to buy in to the tyranny I can throw at them. Defining or chopping people's characteristics into little bits and fixing their meaning is death to the contact we can have. Sometimes the experience of not having contact with the people I love because they don't organise their lives drives me frantically to stop bothering and give up all hope, trying by any means to blackmail them into giving me more, but it never works and frequently destroys the possibility of what would happen if I left it alone. We cannot force love any more than we can force freedom of communication. It is a gift we are more likely to receive if we choose to accept our feelings as generated by our own psychological expectations.

Clearly there are rules in life and relating which we all have some grasp of. When Nick rips off Dot Cotton, or when Peggy Butcher's third marriage heads for disaster we say 'I knew that was going to happen'. Perhaps so many of us watch Eastenders because it gives us life without effort, rather like dreams I've had of getting up in the morning, floating downstairs, doing what has to be done only to notice how easy it all is…because I'm really still dozing in bed. As the audience, we watch Eastenders from God's perspective. Sharon may not know what Dennis is up to, Pauline may not know about Martin and Ian may not know what Laura is doing, but we do. So we watch while holding enough of an overall picture to make a calculated guess as to what's going to happen. How powerful we would be if it were real and we could participate. How often people say of their teenage years, 'If I knew then what I know now.'

In order to create our reality, to bring our lives as closely in line with what we want them to be as we can, we have to grasp the way of things, we have to see our illusions and we have to accept what feels impossible. Perhaps it is even the way of things that the weak create and enable the strong through their hopes and their mistakes. If we grow by a continual process of discarding our assumptions perhaps we even need tyranny for our liberation as we need the dark to see the light and death to become fully conscious of life.

'Dimly, as if recollected from a long-distant past, Albard (the tyrant) sensed his own foolishness. But he was not ashamed. He cared nothing for the mistakes he had made in his life, for he felt now that the Singer people had watched over him, and he had done his duty in his own way. He was ready now for the release.'
Firesong: William Nicholson

If this is the case then we really need to appreciate the strength, the fortitude and the persistence of all our spirits as we go forward with all our frailties.

The Re-Heroising of Our Journey

"What work have I got to do then?" said Will, but went on at once. "No, on second thoughts, don't tell me. I shall decide what I do. If you say my work is fighting, or healing, or exploring, or whatever you might say, I'll always be thinking about it, and if I do end up doing that I'll be resentful because it'll feel as if I didn't have a choice, and if I don't do it, I'll feel guilty because I should. Whatever I do, I will choose it, no one else."
"Then you have already taken the first steps towards wisdom," said Xaphenia.
The Amber Spyglass: Philip Pullman

Clearly the way to begin any long journey, as in all the best fairytales, is to pack enough food for it and to learn how to get more on the way. Once we know how to get and to take in nourishment for our soul we can choose to tune into what is in the way of seeing clearly and discard it. The first thing to go is probably the illusion that there is nothing heroic about going forward. Anyone embracing this life full of losses and heartbreak is heroic. If we are to struggle on, if we are to face our illusions, we need to appreciate this heroism. In *The Lord Of The Rings*, Aragorn is speaking of the trials we all have to face when he tells Arwen what it means to be mortal:

Aragorn:" I speak no comfort to you, for there is no comfort for such pain within the circles of the world. The uttermost choice is before you; to repent and go to the Havens and bear away into the West the memory of our days together that shall there be evergreen but never more than memory; or else to abide the Doom of Men."
Arwen:... "not till now have I understood the tale of your people

and their fall. As wicked fools I scorned them, but I pity them at last. For if this is indeed, as the Eldar say, the gift of the One to Men, it is bitter to receive."
Aragorn: "So it seems ...but let us not be overthrown at the final test, who of old renounced the Shadow and the Ring. In sorrow we must go, but not in despair. Behold! We are not bound for ever to the circles of the world, and beyond them is more than memory, Farewell!"

<u>The Lord of The Rings</u>: *J.R.R.Tolkien*

Culturally we shy away from what is regarded as 'heavy' but the depth of our grief both collectively and independently is nothing to be ashamed of. It lies in our past as well as our future. The darkness of our experience goes beyond what I can communicate. It is necessarily a journey we make alone, an isolation we allow, not knowing where it will lead. In the depths of our psyche lies the very worst of what can happen to us, as infants, as adolescents, as adults, as old people:

"What is the matter with my master?"
asked Sam... "His wound was small, and
it is already closed. There's nothing to be seen
but a cold white mark on his shoulder."
"Frodo has been touched by the weapons of the
enemy...and there is some poison or evil at work
that is beyond my skill to drive out."
<u>The Lord of the Rings</u>: *J.R.R.Tolkien*

There is no healing of our deep, deep wounds – those wounds we can only continue to acknowledge and accept. Deep within us also lies a spark of divinity that knows something is wrong when it sees it.

"They tried to pierce your heart with a
Morgul-knife which remains in the wound.
If they had succeeded, you would have become
like they are, only weaker and under their command."
<u>The Lord of the Rings</u>: *J.R.R.Tolkien*

I speak of those wounds which formed as we formed, as our world formed. Not the cuts received by our hardy knees or even the bones broken when we crashed our bikes. Even when babies are dropped, they bounce; their bones soft and growing can take the knocks. No, I speak here of wounds built into our development along with our food – loss of solidity, an experience of fragmentation, of all that we know breaking up or down. For many of us there are early experiences of such a traumatic nature that we even know what it is to lose things which as adults, we know cannot be lost. Think of a baby, so vulnerable that it cannot even make sense of its experience, needing something that never seems to come. It can't look at the clock and say 'Ok, Mum's only been gone half an hour'. It has no sense of time, no sense that need will be satisfied, no idea of others. What does it make of what, from our perspective, seems like very little trauma? And what about when there is a major one? Mother goes to hospital or is simply unavailable for a long time? This is an experience of nothing as permanent, a nightmare reality so vivid, burnt so deep into our knowing that it is always a possibility to us. Truly this is a journey to the heart of darkness that has already happened. We become filled with feelings so unbearable that we try to make sense of them by directing our fear at something real and tangible, which we can, at least, destroy. Instead of containing and being with our feelings we throw them out in what appears to us to be justified violence but it just generates more fear and despair, like a series of landmines set off by one another's vibrations.

"Every time we open a window with the knife, it makes a Spectre.
It's like a little bit of the abyss that floats out and enters the world."
<u>The Amber Spyglass:</u> *Philip Pullman*

The fear of that possibility can permeate our lives, control our behaviour and hurl us from one illusion and emotion to the next.

On the 13th of that month Farmer Cotton found Frodo lying on his bed.." "It is gone forever," he said, 'and now all is dark and empty'... "I am wounded ...wounded; it will never really heal."
<u>The Lord of the Rings</u>: *J.R.R.Tolkien*

No one would ever come to this planet if they knew what they had to go through and self help or otherwise, real life is not going to be tidied away. Facing it full on. Heroic? What do you think? Living with these wounds means living with the realities we would prefer to ignore, yet in the knowing of them lie the seeds of genuine compassion and our mature soul.

The Art of Magic

> *"Not easy. Not easy at all. Think how much is wanting to make things wrong. All the fear in the world, and the violence that comes from the fear, and the hatred that comes from the violence, and the loneliness that comes from the hatred. All the unhappiness, all the cruelty, it gathers like clouds in the air, and grows dark and cold and heavy, and falls like grey snow in thick layers over the land. Then the world is all muffled and numb and no one can hear each other or feel each other.*
> *Think how sad and lonely that must be."*
> *Firesong: William Nicholson*

Magic is energy used with awareness. If we are aware of what we are experiencing and able to step outside ourselves enough to assess a situation, we can become endlessly discriminating in the way we choose to behave. If we are filled with energy without awareness, we have no magic at our disposal, just a lot of chaos and fear which, if we cannot harness it, will harness us, as well as dragging others into the mess.

At its simplest level, our ability to separate our own experience from what is going on around us means that we can address difficulties from a vantage point that prevents us from taking things personally. If, for instance, someone from Yellow Pages telephones me and starts to tell me how many favours they are doing me, I might feel obliged to buy something. Alternatively I might feel pressurised and angry. By seeing their assumption clearly (that they will try all sorts of manipulation to meet their needs, which is to sell and get commission) I can step out of the muddle and cut through it by merely saying yes (based on my economic needs) or no (based also on my economic needs). In this way, I can minimise the impact they are having on me. One of the qualities we all have is a degree of empathy, but whether we see that for what it is or whether we allow it to drive us crazy is up to us. This book has shown many examples of clients who thought they were mad but are just empathic. We have seen how that energy, interacting with their fear of madness can make them violent in their need to be affirmed as right. There is just too much going on and people panic as they feel themselves sink beneath it. Hanno Hath, father of the central characters in *Firesong* knows these things intuitively:

'For several minutes he climbed the mountain by her side
without speaking, letting her adjust to his nearness. He
believed it was always a mistake to open a conversation
cold; more than a mistake, a kind of assault. It took time,
he felt, for two people to organise their feelings about
each other, to bring them up out of store, before it was
appropriate to speak the first words.'
Firesong: William Nicholson

Being aware of our own feelings and history gives us a sense of ourselves. Simple, but the implications for deeper magic, for what we can do with that, are enormous for if we can separate from our own stuff, we can see what is going on for others much more clearly, and woe betide *them* if they cannot. Potentially, magic gives us power. In the case of Saruman from *The Lord Of The Rings*, he might be a good actor, but his real power comes from the weakness of others, from their desire to conform and not spoil the apparent harmony:

'Suddenly another voice spoke, low and melodious,
its very sound an enchantment. Those who listened
unwarily to that voice could seldom report the words
that they heard; ...mostly they remembered only that
it was a delight to hear the voice speaking, all that it
said seemed wise and reasonable, and desire woke
in them by swift agreement to seem wise themselves.

When others spoke they seemed harsh and uncouth by contrast; and if they gainsaid the voice, anger was kindled in the hearts of those under the spell.
The Lord of the Rings: J.R.R.Tolkien

Once we begin on this track, there is no absolute end. There are hundreds of different realities and the stronger we are in our own, the more able we are to tune into the ones that are occupied by others, and indeed to exploit them if they are unable to hold their own. Magic is a force for bad as well as good, and the fear of those who undertake their heroic journey with another, such as a counsellor, is sensibly based until they have asked the questions and explored the feelings they need to explore to establish that trusting is increasingly appropriate.

If we can establish where we end and others begin, we can leave them to do their own inner sorting of their own difficult feelings. One of the biggest problems the fledgling healer is faced with is their compulsion to make things better. For instance, if someone is in deep denial about something and the fledgling healer has a sense of what is going on at a deeper level, then the fledgling healer will be flooded with a desire to wake the other up. Without really considering what they are doing, they will bang on and on about *they* see as the other's problem while studiously avoiding what is actually going on for *themselves* at a deeper level.

Now this is quite complicated stuff. To make it easier to understand I shall use an example. Let's say that my friend Tom is complaining endlessly about not having a decent job, yet he is educated and fairly able. What *I* might see is his deep fear of, and belief in, his own inadequacy since I have known him for a long time. My initial desire is to wake him up, to make him see that it is his own refusal to commit himself to training, his own fear of not being up to it that is the problem. So, I tell him that. I feel powerful and righteous and if he refuses to face it then that is his look out. He refuses. I never speak to him again and am left feeling virtuous but isolated...now it seems to be my problem. Something is actually not right at all. Nothing has been helped and something has been destroyed.

This is because the whole equation has not been taken into account.

In this exchange, there are two people and the truth is that unless I am willing to use force and thereby disrupt the natural order which will rebound on me later I can only work with, and on, my own energy.

Let us go back to the point at which Tom is endlessly complaining. Yes, my head can see a solution but what is going on at a deeper level *for me?* Oddly enough, an exploration of the first layer shows me that I desperately want to shut him up. It is very, very difficult to listen to this heavy, dark blaming. Aha! That's why I've told him the answer. What I thought was me dispensing wisdom was simply the verbal equivalent of hitting him over the head with a frying pan.

Level two? That is the point at which, if I am willing to look at *myself* and not just *him*, I come into contact with the nature of my *own* pain, the stuff I was trying to avoid by using the frying pan. The truth is that Tom's denial feels as if it is damaging me. I can't stand the fact that he can't see what I can see. I fell as if I am disappearing, as if I don't exist. Ring any bells?

The difference between the fledgling and the mature healer is that the mature healer has done the work, the work in this book, in one form or another. They know it is just a feeling; they are clear about what is their own pain and therefore free to choose how to respond to someone else's. Just because the other is refusing to acknowledge their own struggle doesn't mean it isn't happening, but this is what the fledgling healing fears: that they are imagining things; that they are mad; that their perceptions and therefore their selves have no substance. Tom may or may not decide to acknowledge his feelings. While I am kicking up an endless fuss, that's a distraction for both me and Tom.

To really use your magic is to be able to dance with what is apparent, with what they are *literally* saying while *knowing* at a deeper level what someone else *might* be experiencing. From this point you can address what *might* be needed without forcing anyone to acknowledge anything. You can use that power around what is *unacknowledged.* For instance, I might say 'so, no one giving you a decent job must have a real impact on your morale' and give him a different and shame free route to connecting with a deeper level of his feeling.

It is also crucial to have enough humility to know that we might be wrong. We never know anything absolutely, and we certainly don't know the whole picture. So can you hold yourself and all your frustration in abeyance? Always we come back to the quest.

'He had supposed effort was needed not to fall. He had been straining invisible muscles. But no effort was required. He needed stillness. It was more like finding the point of perfect balance when standing on one leg. Until you find it, you wobble and wave your arms, but once you have it, perfect stillness keeps you in place.'
Firesong: William Nicholson

When we talk of using magic we are always referring to occupying a minimum of two realities. For example, consider a situation where one person feels rather distant. Three realities around that might be:

- The literal reality of the other: 'No dear I'm just tired'
- The deeper reality of our knowing: 'He/she is avoiding how angry he/she is about …"
- Our own feelings of lack of contact, alienation, not being good enough etc.

Unless we can hold, see and accept all three, it becomes impossible for us to 'dance' with them and by dancing I mean addressing them separately by *choosing* the most appropriate and effective response for us *and* them:

- 'Why not go to bed now then'
- Is this a reasonable anger? Perhaps I should wash up, cook and leave a mess or be ready to say exactly the wrong thing in the morning and clear the air.
- I think I'll go and see some friends.

Trying to deal with the unconscious using rational and reasonable means will only cause disturbance.

'How maddening her brother could be, with his assumption that he knew her better than she knew herself.'
Firesong: William Nicholson

This is hardly magic!

We live always within a whirling sea of existential insecurity. There is no truth in the idea that even our closest relationships are secure. It is nonsense to say "If only we ...." (Please fill in the dots with your own personal delusion!) everything will be OK. and sometimes things are quite simple:

> *'The baby and the cowman between them somehow put her into a better humour. She* (Pinto*) liked it that they had their own concerns, that didn't include her.'*
> Firesong: *William Nicholson*

Between casting off illusions and accepting that things change there is the possibility of being with what *is.* Take love. The secret of being loved is particularly difficult to recognise because it is not earth shattering but subtle. It is not 'just something that happens' but takes discipline and awareness even on the part of the beloved. Love is not a yearning to be complete or having all your needs met. It is more likely to be found in accepting and trusting that somebody (appropriate) loves you. Rather like needing the chink in the wall to observe me beating myself up, I need a chink in the wall to notice when someone loves me too. Like the *Sleeping Beauty* I can wake to the man who responds and gives love and equally, I can doze with one eye open when the love begins to fade and I can behave in a way that is appropriate to nurturing the love and respect for myself. It's all back to the dance. Being responsive to how things *are*. I have to constantly be ready to move towards people but equally, I have to be ready to move away from them. This is true power and holding it is difficult. Sometimes you can tell people secrets or magic or truth and your insight will be weakened because it will be received badly, perhaps disrespected or not heard. Others may transform insight into cliché and the power is thrown away. We all know that a wife may nag and nag and her insights might be spot on but to no effect. Others, when the timing is right (and this is particularly true in the counselling room) may fully hear something and be transformed. Then a reality can become strengthened and more powerful. The magic ingredient changes every time – the clue to it lies in our intuitive ability to assess that situation and to access real contact, meeting others where they are.

Several months after the death of my father I dreamed he was alive, while knowing somehow that he was dead. In the dream I struggled with this apparent contradiction. I was filled with love for him. Love far greater than I had ever felt while he was alive and I asked him, over and over: 'How can this be?' I had seen him die; sat with his body; buried him. He was typically him in the dream, fixing me with a look that contained a sigh of slight frustration as he sought to find an answer that would make sense. And then he spoke. He said: 'Well Gillian' (he always called me that) 'You have a great deal of love to give.' I woke up knowing that the rational answer to my dilemma was that I had been dreaming. But there was a deeper knowing that truth lies not in the facts but in the depth of contact between people, just as it lies in the depth of contact with oneself and other parts of oneself, some of which are anything but rational or very earthly. We have plenty of portkeys, this one being sleep, others being fireplaces, candles, stone circles, places of beauty, which are really altars transporting us through doors to other ways of seeing. If we treat them with the respect they deserve they can provide a focus and access to parts of ourselves and our perception, call it our soul, our spirit, our hopes, our gods, our higher self, the dead, which enable us to step into different realities.

It is true that you can't get away from the bad, only move through it, but it is also true that you can't get away from the good. In another dream my grief was so great that I couldn't scream loudly enough. I ran out into the landscape where it seemed that the deep comfort many of us associate with home resided. Once outside, the contours of the land held me like a baby. It seemed that the rivers flowed, not with, but because of the tears of the Goddess/ the mystery/ the great being. They were there for me to pour my grief into just as my parents once were, or weren't, and I felt that, putting away childish things, it was not my parent's images I wanted to be in. They are pale and transient imitations of a deeper reality. I discovered that it's not the spirit of people you find in beautiful places, it is the spirit of the place that was always in those people, and that seems to go on.

The point about death is that we don't know anything about it. Perhaps it is this that is so frightening. If we have vivid imaginations we can sometimes use what we *do* know to frighten ourselves silly. T.S.Eliot conjured up a particularly vivid horror in *East Coker*, part of his long poem *The Four Quartets.* Here he writes about darkness and the endless vacant spaces that everything dissolves into; where everything we love we lose; where form itself unravels into emptiness. But T.S.Eliot is writing of fear; loneliness; of the horrors at the heart of darkness *that are part of life*, rather than death itself. We need to see this knowing for what it is rather than as a fixed belief or terror that we shrink from and try to ignore. We need to know the very dark stuff so we can choose *not* to step into it, as well as *when* to step into it. We are free within limits. We have choices. We can empathise in our frailty, our vulnerability, our difficult feelings and we can treat others as we would choose to treat ourselves, our whole selves. When we know a feeling fully, we can move away and if we're less bound up with being good in relation to our past scripts or with being fixed in one reality, then we can choose our immediate future and probably change our long term future.

"I will not give you counsel, saying do this, or do that.
For not in doing or contriving, nor in choosing between
this course and another can I avail; but only in knowing
what was and is, and in part also what shall be."
Gandalf in The Lord of the Rings: J.R.R.Tolkien

Exercise

Changing Consciousness At Will:

For this you will need your box and all your messages and the safe space that, by now has become second nature to take and to hold onto. If it hasn't, you're doing the wrong exercise.

Tip all your paper scrolls out of the box and onto the floor. Remember all that redness at the beginning? All the harshness? What softness surrounds them now? A baby pink or blue or pale yellow perhaps. Something soothing and calming.

So what's to be said? If you have made your way from unconsciousness to a deep awareness of the messages you give yourself ...If you have returned those messages to their source and replaced them with a more nurturing response to what is difficult, frustrating or sad then you have learned magic because you have learned to change your consciousness at will.

Look at your scrolls of paper on the floor. Look at the wisdom and insight they represent. Look at the aspects of yourself you have accessed, look at the pain you have tolerated and the care you have chosen to give yourself. Every single time you even acknowledge this, you change your consciousness.

Loosening the flow

1. Allow the thoughts in your head to float by. Watch them. Perhaps they are quite familiar, perhaps they are questions. Without becoming too attached to them notice how you feel but instead of attaching a label to it, try to locate where it is within your body. See it as an image or a colour.
2. Notice how quickly the thoughts gather round and attach themselves to this feeling, as though you are trying to make sense of it.
3. Perhaps you see this image as a red ball, flaming with anger or as a black knot that is tangled and stuck or rain drenching you clean. Allow the image to detach from the meaning/the words/ the thoughts and follow it in your mind's eye. Watch it as it moves further and further from these constructed thoughts and beliefs.
4. Like a rolling stone gathering no moss, let all the thoughts be left behind in the world of constructed meaning and keep watching as the image and the feelings transform and transmute into something else. What does it need to do to sort itself out and have a break?

There are millions of realities out/in there. Having the ability to loosen the grip of the one we most compulsively jump into gives us a huge range of possibilities regarding our choices and our capacity to contribute to the world and ourselves.

The next time you find yourself falling into a response that is familiar, take a step back from your emotion, imagine the feeling and detach it from the meaning. Then consider, if you are not the person who always does/feels/believes a particular thing, who the hell are you?

I think you must be the guy watching the show; stepping in and out of your personality like it's your favourite jumper, putting it through the wash, dying it another colour, patching it, incorporating it into a bedspread because you love it too much to chuck it out completely. And why not? If it unravels one day, well it's only a jumper. It's not the end of the world.

It would seem that we are freer than we could have imagined and we don't need to live in a fantasy in order to know it. We are forged of too much pain to simply be lost. All we need is the spontaneity of our felt sense and the courage to fully live.

Chapter Ten

What Is True?

"Your reading will be even better then, after a lifetime of thought and effort, because it will come from conscious understanding. Grace attained like that is deeper and fuller than grace that comes freely and furthermore, once you've gained it, it will never leave you."
The Amber Spyglass: *Philip Pullman*

The Self as Reference Point

Knowing what we feel means we can constantly refer to it when everything else is in doubt. It makes it possible to stand our ground and to speak our truth against opposition or when we are being ignored. It precludes the need to avoid confrontation because we no longer fear the consequences of our feelings. They are, after all, only feelings. We will not *actually* die or collapse or disappear. These are fears. Once we know what we feel we can see these feelings for what they are. Once we have awareness, we have choice. Not being aware of what we feel in our being means we are hostage to the vagaries of our families, our societies or our own intellectual opinions and when these change we have no signposts telling us what is in our best interests. At its most basic level this occurs when we suffer and are unable even to realise that we are hungry. On a more sophisticated level we may rage at people and blame them because we have not 'tuned in' to realise that we are anxious about something. Awareness leads to choice. Will power works both ways: to forge ahead ignoring our needs or to give in and attend to what is important in ourselves. We are free if we choose to be, if we choose to tune in to our needs. We are also lucky enough to live in a society whose democratic values are enshrined in law. A victim of torture in Chile may be entirely enlightened and still be unable to escape from external circumstance but for us, in the West, we have the right to fight against this enshrined in law. Under these circumstances a disproportionate part of what happens to us as adults happens because we unconsciously collude with an idea and in this way, we do it to ourselves. Looked at in this way, even the worst part of our experience is ultimately liberating because that which we make we can undo.

The degree to how much we are disconnected from why we do what we do can be quite shocking. What is it that makes us tick? What makes us feel good or bad? Madelaine Albright was former Ambassador to the UN and U.S. Secretary of State under Bill Clinton. Brought up Catholic, she fled the Nazis as a very young child, discovering as an adult that she had been born Jewish and lost several members of her family in the Holocaust, yet she is frequently asked the question "Do you think your childhood has affected your career choice?" I wonder if you are surprised at how often she has to answer this or if you also think perhaps her decision to devote her life to trying to negotiate peace between hostile nations was a complete coincidence?

We learn what we think we know because of the way things make sense to us in relationship. Meaning becomes strengthened because it is shared and identified, even if we are deluded. Meaning is alive, moving, changing and formed of patterns viewed from perspectives. They are partial and precarious; almost tangible and often temporary. They generate realities that stop our minds from being blown apart. The question is: 'Whose perspective?' How much of a perspective is our own and how much is someone else's? As a child I was always closer to my mother, who was soft and sympathetic, than to my father, who was forbidding, exacting and sometimes downright bullying. It was easy to adopt her perspective of the way things were in the family because I felt more dependent on her, was closer to her and so saw things through her eyes. Until he died! It is true that all through my life I was told I was like him, but it would seem that empathy and the unbearable reality of his being extinguished somehow moved me further into his shoes than I had ever been. All of his irritation at my mother became suddenly absolutely clear, and all that he had sniped about her, which was far less than my mother had said behind *his* back, gave me a very startling picture of *her* shortcomings. Their constant and even unspoken blaming of one another for their unhappiness had been very confusing. Eventually I was left with a panoramic view of both of them, including their blindspots: my point of view as an adult at last. I called it Total Perspective Vortex. For months and months, I had tried to dovetail them into a harmonious picture, oscillating between one perspective and another, until one day I realised that *they* never dovetailed, they never worked it out. They just battered away from their own perspectives. I have since wondered whether their warring way encouraged me to compartmentalise my life, different parts of me being satisfied by a multiplicity of affairs with very different people rather than becoming more integrated by sharing all of myself with one. The vulnerability needed to do this was never demonstrated, the only way seemed to be a stultifying compromise where parts of them were not expressed to one another, the kind of life I was not willing to embrace. Perhaps we come to real compromise, real surrender, through fully expressing and working through that conflict. Perhaps the magic ingredient is in beginning to see what is important, rather than maintaining a defensive obsession with what's fair or right, and we seem to soften in the way that steaks get tenderised – from hard, almost unbearable knocks to do with the truth about reality, death, solitude, and loss.

'Like a clay pot that has to be repeatedly fired in order to withstand the pressure of what it is to hold, so we are tempered in preparation for the future.'
Where Two Worlds Touch: *Gloria Karpinski*

Perhaps it is not so very inappropriate that just as we find the identities we have fought so hard to develop, everything changes yet again and we begin to forget what was so very important about them.

"That's the trouble with life," said Mist.
"In the end, you lose interest."
Firesong: *William Nicholson*

Since we must die, that sounds quite fitting.

While we live though, we constantly sort the dross from the gold through our felt sense and today's gold is necessarily tomorrow's dross. That's why our felt sense is so important. It is sifter, compass and creator. Without it we are not ourselves and occupy an alien and utterly insecure medium of meaninglessness. The word *thankyou* means nothing without a deep appreciation of what has been done for you, just as the word *sorry* has to understand the level of injury to not be hollow. The word *love*, also, has to be tapped at its source.

"If you make my brother happy, you make me happy. If he loves you,
I love you. If you love him, you love me ...my other myself."
Firesong: William Nicholson

The Miracle of Acceptance

As we learn to live with the different experiences we've had, the pain subsides and we live increasingly in the now. The more pain we learn to live with the more aware we become of the pain we haven't got. Instead of fighting our distress we accept it and accepting it, we follow where it takes us:

'In desperate hope I go and search for her
in all the corners of my room; I find her not.
My house is small and what once has
gone from it can never be regained.
But infinite is thy mansion, my lord, and
seeking her I have come to thy door.
I stand under the golden canopy of thine evening
sky and I lift my eager eyes to thy face..
I have come to the brink of eternity from which
nothing can vanish – no hope, no happiness,
no vision of a face seen through tears.
Oh dip my emptied life into that ocean, plunge it
into the deepest fullness. Let me for once feel that
lost sweet touch in the allness of the universe.'
Rabindranath Tagore: Gitanjali LXXXV11

As we begin to accept the reality of death, with the loss that is permanent, we move from hope to resignation. Illusions are felled but we continue to feel what must be felt. Since all feeling is life and life is movement, the search goes on despite our apathy. All that we intended has not come to be. That which I wanted to change has not all been changed in tune with my attempts.

As I accept certain immutable facts, my perspective expands to accommodate what might have been unthinkable fifteen years ago. That which I found shocking or unacceptable about myself once and tried to hide or change, I no longer think so awful or so important. Perhaps my desires have become less central as I know more about ageing and death. As Bowman discovers with utter astonishment when he begins to see his own shortcomings:

'His destiny was not to be the saviour, but to be saved.'
Firesong: *William Nicholson*

He is not at the centre of things at all, but part of something much larger. In *Slaves Of The Mastery,* the second part of *The Wind On Fire* trilogy, Bowman and Kestrel come face to face with supreme evil – the Morah, but it is not a traditional evil – not the bad guy with the horns. The Morah is inside us all. It is the spirit of Hitler's Germany, of youth-led persecution. The feeling of strength and invulnerability achieved by surrendering to the Morah is almost irresistible. By its nature, it loves only to kill. Its motto is:

'No more fear...let others fear now. All were young, all were beautiful and all were smiling.'
Slaves Of The Mastery: *William Nicholson*

In this surrender to the majority will there is no conflict, no opposition and no consciousness beyond the persecution of others. There is nothing but a loveless and ruthless lust for power. By now, we should not be surprised that it is his love for Kestrel, and hers for him, the recognition of his own sweet care and his own sweet vulnerability, that reconnects and redeems him. And then he is left with an extraordinary gift. His confrontation with the Morah has given him experience of every facet of human nature and feeling. He understands what it is to be 'bad' as much as what it is to be 'good'. Bowman's perspective, his deep inside knowledge of people's hopes and needs and fears, of their kindness and their cruelty, means that he can see these things not in conflict with one another but simply for what they are. He has become a meeting place for them.

'Men killed and burned not for gain, not for power,
not even for pleasure, but because they craved destruction.
They had lost everything. Now they were determined
that everything should be lost.'
Firesong: William Nicholson

In our desire to distance ourselves from these unpalatable truths, we judge them and so cannot enter their reality. We reject them rather than accepting them as true and by rejecting them, we keep them stuck – endlessly denying them in ourselves, constantly criticising them in others. If we simply accept the truth of them, and know this truth, perhaps something will change:

'He wept for those who suffered and for those, almost
equally helpless, who inflicted the suffering.
"I understand you all," he called down to them.
"I am the guilty one who will be saved for you and with
you, so that the world may start again."
Firesong: William Nicholson

For Bowman, this is where acceptance leads – not to an emotional judgement and a stagnation, but to a wider perspective, a stepping outside of what he can see so that he can see more, can see it from a wider angle and so become different.

There seems to always be this need in our stories. Contradictions have to be present if transcendence is to occur. In order for us to be saved, someone or something has to hold the notion of saving. In the end they are equal, needing one another in order for movement to happen. In *Firesong,* we are led to believe, as is Bowman, that he is the saviour. Perhaps this too is necessary in order for him to access the last dregs of his desire to be important and realise that they too are illusion so that he can let them go and allow the process of transformation to take over.

'Kestrel heard all the anger and fear, the envy and the mockery, the hatred of happiness and hunger for pain. (She) saw the contorted faces of the people, and heard their sobbing cries and she knew, this is me, this is us, this is mankind. This was the world of hurt and loss that must die, for the time of kindness to come again.'
Firesong: William Nicholson

Finally it is Kestrel, his sister who walks into the flame:

'Here, now, lost in a greater light, she knew there was no end, no boundaries, no this and that, no then and now. All existence had melted, including the thing she had once known as her own body – and my mind? My self? Is that melted away?'
Firesong: William Nicholson

The point is that it doesn't really matter.

The Paradox of Death

"I have been too deeply hurt, Sam. I tried to save the Shire, and it has been saved, but not for me. It must often be so Sam, when things are in danger: some one has to give them up, lose them, so that others may keep them"
The Lord of the Rings: J.R.R.Tolkien

What is completely broken is completely broken. It cannot simply be mended though often it seems that the lessons passed on go towards a greater mending. Sometimes there is no healing from the worst of our experience, there is only respite. Some of the worst places we have been are unspeakable in their horror. The death or loss for reasons beyond our control of those we love remains a unique agony, yet we all experience this. At some point in time we all suffer this profound tearing, as if our souls are being ripped out.

At the time this is so all consuming it seems as if there is no other reality, but of course there is – what else would we be mourning for?

"Go in peace! I will not say: do not weep;
for not all tears are an evil."
<u>*The Lord of the Rings*</u>*: J.R.R.Tolkien*

When love appears to abandon us, what is left? Frodo has been touched by the morghul knife wielded by the Nazgul, wounded beyond any healing skills that the Shire or even Gandalf can provide. It is no coincidence that the time of those who held the wisdom and the power is over. They too had rings: Galadriel had Nenya, Elrond had Vilya and Gandalf, Narya. While Sauron's great ring existed, so did these, to balance good and evil. At the end of *The Lord Of The Rings* the ring of evil is destroyed, cast into the fires of Mount Doom. The One Ring to bind them to Sauron's will is gone. Middle Earth is free, but freedom is more than a release from a dominating evil. Up to this point, Galadriel, Elrond and Gandalf have been like Gods and the men and hobbits like children: innocent, not responsible for themselves but in their leaving, the time that Tolkien calls 'The Third Age' is over; the time of conveniently handing responsibility to someone who knows better is past.

The hobbits are closer to the consequences of their choices than they have ever been before. Peril is dangerously near but there is an upside to the horrors they have witnessed:

"People will remember the great danger
and so love their beloved land all the more."
<u>*The Lord of the Rings*</u>*: J.R.R.Tolkien*

In Philip Pullman's trilogy the same point is made when he talks about the church as 'The Authority'. 'The Authority' is simply a dogma of received ideas. It allows no questioning or exploration. Its inflexible rules restrict our freedom to make our own decisions and be fully alive. Who needs yet another 'Authority' when our lives are already restricted, contained and bounded by the limits of being human?

"D'you remember another thing he said, my
father? He said we have to build the republic of

heaven where we are. He said that
for us there isn't anywhere else."
The Amber Spyglass: *Philip Pullman*

Death and separation seem to enable us to plumb the depths of our existence like nothing else. The paradox certainly seems to be that we can't grasp anything of real value until we've lost it; that being human we learn the hard way over and over again and that far from death and separation being the end, they are absolutely necessary to generate consciousness. At the end of *The Amber Spyglass,* Will and Lyra are faced with a choice. Because they come from different worlds they cannot co-exist. One of them will deteriorate and die very quickly. In *The Subtle Knife*, the second book of the trilogy, they inherited a blade that could cut the boundary between the many different realities, or worlds. This is how they travelled together. By the end of the trilogy it is apparent that so many windows have been left open between worlds (not just by them) that the dust – the very essence of consciousness, is leaking so fast into the great void that all the windows must be closed never to be opened again. They are faced either with separating forever or with being sucked into nothing:

"I wish this night would never end! If only we could
stay here like this, and the world could stop turning, and
everyone else could fall into a sleep.... I'll be looking for
you, Will, every moment, every single moment. And when
we do find each other again we'll cling together so tight
that nothing and no one'll ever tear us apart.
Every atom of me and every atom of you"...
The Amber Spyglass: *Philip Pullman*

This searching is what will make Lyra grow and be wise, keeping her eye on the truth she knows to be real; a wholeness that cannot be broken, unless she chooses to break it. She may be separated, but she knows of these other worlds. It is the same truth that keeps Buffy moving on when she has lost Angel, the archetypal Romeo that she too had to lose in order for them both to live.

"There comes a time in partings
when all that is left to be done is to part."
Firesong: *William Nicholson*

Again and again we find that separation generates consciousness: Renton leaving Scotland, Maria leaving the nunnery, Harry Potter losing his parents. Can it be that death is as essential to this human life as food and love?

At the start of this chapter I referred to our coming of age. Our age. The one where the Gods are seen for what they are. Not punitive or narrow, just a pure awareness of how things work for humans. If we talk about the will of God, that often sets off more assumptions to wade through, but if we talk about the will of Love, perhaps that is more helpful. It seems that we do work in a certain way and love is always available to see us through, much like the good parent. All we have to do is be as fully connected to ourselves as we can be because, if there is something else, it's not for us to know yet.

"You said 'You're going to die soon'. Don't you see
how easy that makes everything?...Everything you
might ever have done in all your life has to
be done now, or you'll never do it."
Firesong: *William Nicholson*

Our reality is that death puts us in touch with how much we love and how much we care like nothing else does. It puts us in touch with knowing who we are and what we want. For instance, as you might have noticed by now, I am a person who likes to make huge sweeping statements about things. Perhaps that is not so dreadful. If I can accept that without condemning it, it will change of its own accord. It will right and rebalance itself. There is always this duality, this paradox. As I know myself more, I step further from my idea of myself. As we break down and let go of our patterns of meaning we move closer to Kestrel, as she enters the flame:

'There was only this infinity of humming
light – and herself, knowing it.'
Firesong: *William Nicholson*

Staying with what is, with what is commonplace, we find paradoxically that it is extraordinary. If we take the wide-angle lens approach, imagining awareness as infinite, like the universe, where does this take us? Initially, perspective draws back. If I look at my father in a wider context, he becomes 'Tony' because he is seen by a very different audience: he becomes no longer my very personalised dad but becomes grandfather, distant cousin, ancestor. What if the lens keeps pulling out more and more?

Perhaps it is time to tell ourselves some stories beyond our limits: the logical conclusion to some of the healthier assumptions I have asserted in this book, conclusions grounded in our wider knowing, in our experiential knowing. For instance:

- If we express our experience without censoring it, we will move through it and change

The universe is expanding, the planets becoming further and further from one another. This much we know. Like a photograph of ourselves magnified a million times, the dots that make it up become increasingly separated, just molecules, no picture, no judgement.

'She poured herself into the flame, and
the more she emptied, the lighter she
became, and the more she was swept and
scoured by joy. Knowing the end was very
close, she cried out through her song.
"I love you Bo! I love you all!
All my loves..."
The last Kestrel felt was a wild singing,
a white-hot melting glory. She was
falling into the light...lost in bliss.'
Firesong: *William Nicholson*

And what of Bo, left behind, still that unique constellation of himself as human being?

'The form of his sister still stood before him, his hands could hold her, but she herself, all that he knew of her, was gone...He found her then: so close that he couldn't see her or touch her or sense her in any way, except as part of what he saw and touched and sensed for himself...He looked at her, smiling before him and saw her dear face: only it was his own face he saw, and he was seeing with her eyes.'
Firesong: William Nicholson

He is left with his awareness, of her; he is left with knowing there is eternal love when the rest is burned away.

Perhaps stories are better than theory, and who is to say that this story is worse that any other? If we start with some of our new assumptions and follow the logic that unfolds from our experience into the ever expanding universe, perhaps we come closer to touching a wholeness than if we reduce, overdefine and limit.

I would like this book to end like all the best stories, like Hans Christian Anderson's *Bluebird of Happiness.* In this fairytale, brother and sister, Mytle and Tytle go on a long quest to find their mythical saviour only to discover that it is the commonplace bird always in their back garden. They, like us, have searched for life's meaning only to find that they have created their own reality. The truth is that the meaning was there all the time hidden beneath layers of illusion that they have generated.

The secret is knowing that it's not at all easy, not all noble or tragic or joyful. Some of the spadework is quite dull – which, paradoxically is probably why we left home in the first place.

In *The Last Battle*, C.S.Lewis's final book about Narnia, the land has become so corrupt that the battle is lost, Narnia is finished and the children we have followed so closely, and that have followed their own hearts so loyally, are dead, yet they find themselves somewhere strangely familiar:

"Could it be somewhere we once stayed for a holiday when we were very, very small?"...
"I bet there isn't a country like this anywhere in our world. Look at those colours! You couldn't get a blue like the blue on those mountains in our world."...
"If you ask me, said Edmund, "it's like somewhere in the Narnian world. Look at those mountains ahead...Surely they're rather like the mountains we used to see from Narnia, the ones up Westward beyond the Waterfall?"
"Yes, so they are," said Peter. "Only these are bigger."...
"Those hills," said Lucy, "the nice woody ones and the blue ones behind – aren't they very like the Southern borders of Narnia?"
"Like!" cried Edmund after a moment's silence. "Why, they're exactly like."
"And yet they're not like," said Lucy. "They're different. They have more colours on them and they look further away than I remembered and they're more... more...oh, I don't know..."
"More like the real thing," said the Lord Digory softly.
The Last Battle: C.S.Lewis

In the way that falling in love is a time when we know, sense, touch something of our essence, something hidden from us beneath the everyday, the children have found themselves in a land where what is really important, where what truly matters, has come closer. As we have seen, we are endlessly preoccupied with ideas that seem essential, but we all have those moments (usually at a death or birth or when we fall in love) when the veil lifts and we realise our deception. Our perception is so restricted by our inability to see beyond ourselves as the epicentre of existence that we only catch a glimpse of how things are when we are knocked out ourselves, leaving behind our own self generated debris.

"When Aslan said you could never go back to Narnia, he meant the Narnia you were thinking of. But that was not the real Narnia. That had a beginning and an end. It was only a shadow or a copy of the real Narnia which has always been here and will always be here...All of the

old Narnia that mattered, all the dear creatures, have been drawn into the real Narnia through the door. And of course it is different, as different as a real thing is from a shadow or as waking life is from a dream."
The Last Battle: *C.S.Lewis*

We all know how our personal debris – the piled up disappointments, the atrophied trauma, the cynical beliefs - can make us turn from the love on offer. The dwarves too have lost the battle and are in this new land. It is filled with the juiciest fruit, the most thirst quenching water, but the dwarves, who have also gone through the door, believe so firmly that they have only gone into a dirty stable that they are unable to see what else is going on.

'(Lucy) leant across and held the fresh, damp flowers to Diggle's ugly nose..."How dare you! What do you mean by shoving a lot of filthy stable litter in my face"...
Aslan raised his head and shook his mane. Instantly a glorious feast appeared...but...they thought they were eating and drinking only the sort of things you might find in a stable..."Ugh! Fancy drinking dirty water out of a trough that a donkey's been at."
"You see," said Aslan. "They will not let us help them...Their prison is only in their own minds, yet they are in that prison; and so afraid of being taken in that they cannot be taken out."'
The Last Battle: *C.S.Lewis*

If the universe is expanding, it stands to reason that there is always more, something bigger. Big enough to contain as many small minds as choose to fill it. And if you choose to be big minded?

"The further up and the further in you go, the bigger everything gets. The inside is larger than the outside."
Lucy looked hard at the garden and saw that it was not really a garden but a whole world, with its own rivers and woods and sea and mountains. But they were not strange: she knew them all.
"I see," she said. "This is still Narnia, and more real and more

beautiful than the Narnia outside the stable door! I see ... world within world, Narnia within Narnia..."
"Yes," said Mr Tumnus, "like an onion: except that as you go in and in, each circle is larger than the last."
<u>The Last Battle:</u> *C.S.Lewis*

Knowing ourselves – knowing our not-knowing – reaching for our dreams – dreams which seem so similar for humans across the world; dreams that beat in every unfettered heart. Surely this is a reality that we can choose to come to know:

"This is the land I have been looking for all
my life, though I never knew it till now.
The reason why we loved the old Narnia is
that it sometimes looked a little like this."
<u>The Last Battle:</u> *C.S.Lewis*

It is:

'-deeper, more wonderful, more like places in a story:
in a story you have never heard but very much
want to know...every rock and flower and blade
of grass looked as if it meant more.'
<u>The Last Battle:</u> *C.S.Lewis*

Perhaps it meant more because the unimportant stuff wasn't clouding their vision any longer. No desire, no yearning, no ill frail bodies, no restricted time and space – humanness all done away with as all things must pass – but as Bo says, in his deep connection with his twin, where briefly, although human, he touches the pure fire, where the dross is burned away:

"Why should it ever end?
Why shouldn't we
love each other for ever?"
<u>Firesong:</u> *William Nicholson*

Why indeed?

APPENDIX

Harlow:

Nothing to do with the twenties screen Goddess Jean, this Harlow used monkeys to try and find out what was most important to the young: food or comfort? He set up a series of dummy mother monkeys, some made of wire mesh, some with no heads, and some of soft and cuddly terry towelling. Even without food the choice was clear, the cuddly mum won every time. When he placed a bottle containing milk in the wire mesh dummy, the baby monkeys went over to it only to feed and even showed signs of fear and dislike before running back to the comfort of the soft, though foodless, mother substitute.

Harlow used monkeys because experiments to see the effects of emotional deprivation on humans would be unethical. You simply cannot lock a child in a room with an unresponsive or ambivalent mother to see if it gets scared and becomes withdrawn. What you can do, though, is observe children and find out what effect Mum's behaviour has long term.

Bowlby:

John Bowlby was very down to earth. What he didn't observe he declined to comment on. He would have been horrified at our ability to blame our problems on this or that separation from mum at a crucial time, seemingly developing a habit of reducing our current difficulties in relationships to some source as if that can solve them. Bowlby started from the other end, with babies, observing them as they grew up and noting every separation, how responsive the mother was to the child and the behaviour of the children. In scientific psychology, it is nearly always *behaviour* that is measured for the simple reason that feelings and subjective experience cannot be observed by someone outside oneself. Bowlby observed the behaviour of children for long enough and with *enough* children to be able to state unequivocally that: how available and responsive the main caregiver was, made a significant difference to how the child then behaved and developed.

If you were in a boat on the sea and wanted to go for a swim, you would anchor your boat so that it was there for you to return to. Young children are similar in their need, constantly checking whether the caregiver is still there and available. Will she/he respond? Once they know that she/he will, they wander off again, content and secure. At some point they will check again. In a way, you could say that the young children are checking on what anchors them. My own daughter does this still on her return from school when she comes in, slams the door, shouts 'Hi Mum' and waits for a response. Once she gets it, off she goes and I will probably not see her until she demands food.

What Bowlby observed was that if the main parent was *responsive* and there to rely on when the child needed them, the child was secure and grew into a person who was:

- Co operative
- Resilient
- Resourceful

It would seem that the secure base, the anchored caregiver, always available to help (or ready to explain why not, once a child is old enough) meant that the child became more able to cope with life's demands. Studies showed that these children could be flexible and open to new ideas and new ways of looking at the world and themselves. They could grow and change with experience. If we take the boat metaphor further it is as if, eventually, the child swam away knowing that they would never drown, knowing that seeing the world differently would never make them forget the boat. No matter where they go, the boat is always with them. They were so secure in themselves that they didn't even have to remember it!

How different that is from a sense of 'I must remember the way I came otherwise I'll never be able to get back.' Bowlby found that *unresponsive* mothers who were not there when the child needed them produced children who constantly sought attention. They needed their boat and they were always looking for it, over and over again. While this was going on, everything else went on hold, so whereas the secure child could relax and learn with excitement, the anxious child clung to what they knew (and one of the things they knew was that it was an unsafe world). Emotionally therefore, they couldn't grow up. They couldn't take in new information which would help them see themselves and their world differently, but chose instead to turn away from it, searching only for the boat that they believed would make them feel everything was ok. These *anxious* children fell into two types: *Anxious avoidant*, who were:

- Insulated
- Hostile
- Anti-social

and *Anxious Resistant*, who were:

- Tense
- Impulsive
- Impatient
- Passive/Helpless

At this point in the chapter I can only say try not to worry. Disturbed patterns of attachment are endemic in this country, though they seem to be less so in very different cultures where children are carried within small communities until they choose to be put down.

What Bowlby found then, was that the development of the emotional life was dependent on the bonds between the child and the main carer (usually the mother). Specifically the emotional life depended on whether those bonds were solidly formed, and whether they were consistently maintained or disrupted.

As children develop and become mobile they move between exploring the surroundings and checking that mother is still around. The mother intervenes to identify potential danger. Slowly the child builds a realistic confidence by a method of trial and error, having any major mishaps made OK by the care and the healing lavished by the main carer. If the main carer disappears when really needed (as much for reassurance as to take total control) this is described (within psychology) as loss of the attachment figure, and leads to behaviour that is principally motivated by fear of that being repeated. This behaviour is mainly an attempt to *avoid* the emotions related to this negative experience. That is, loss or threat of loss of the main carer sets off difficult feelings and behaviour, particularly if we are young enough and vulnerable enough for these to seem life threatening; often these feelings might be a strong sense of not knowing if things will work out OK or not. *If not worked through, this avoiding behaviour is something that we continue to repeat into later life and carry into our main relationships*. Thus the punishment we give to partners who set off these feelings perhaps by not being there when we want them, rarely fits the crime because it is a punishment aimed at an earlier time to do with a root, and often, very extreme experience.

Our stability in early attachment relationships is connected to our capacity to love and be loved in later life: how aggressive we are; our attitude to our friends, our teachers and in particular, authority. Crime, delinquency and anti-social behaviour can often be traced back to an experience of potentially violent parents with the mother frequently threatening to leave in order to discipline the child. This leads to anxiety and anger which cannot be expressed in case it really does lead to the parent leaving, thus it goes underground and pops up to wreak havoc on those more vulnerable or accessible, or indeed on ourselves. In particular, those children that are not *secure* (see above) are far more likely either to be bullied (*anxious resistant*) or to bully others (*anxious avoidant*). (Sroufe 1988)

The baby is learning how to communicate all the time. It has a series of ways to behave: it smiles and cries, follows, looks at, approaches, all of which aim to bring the mother closer. What type of behaviour is likely to achieve this end result depends on what makes the main caregiver respond. As we grow, we become better and better at getting the desired response. As we grow older, we form different bonds with all those we are involved with, though it would seem that early disruption often leaves us unable to see others clearly and more or less stuck in one way of being. The bigger the disruption/s the greater the inability to see the others for who they really are.

The work of Bowlby highlighted and proved that meeting the needs of young children had a direct impact on the way in which they developed. Continuity, reassurance and recognition of concerns and feelings enabled children to make sense of their experience in a reasonably accurate way, and built a secure base to grow from.

Piaget:

Piaget studied children's cognitive development. (that is, their ability to think in increasingly sophisticated ways.) What he found, like all the others, was that children could not go from step 1 to step 6 without going 2, 3, 4 and 5 first. Those who were forced in their learning tended to develop sophisticated ways of hiding the fact that they didn't know what they were talking about. The stages he described were:

- Sensorimotor: Where automatic reflexes like sucking are built on and added to until the child can link them together to do more complex things such as stand and walk.
- Preoperational: Where children develop language and so have internal representations of things. They are only aware of other people and things in terms of themselves and what they want or need.
- Concrete Operational: Where children can get the hang of how others see things and can understand the complexity of a concrete (practical) problem and how to resolve it. They are also able to grasp that things change instead of staying the same.
- Formal Operational: Where children can think of things in abstract terms and consider things logically and hypothetically. The nature of freedom, justice, reality and the possibility that everything I have said up to now is just one way of looking at the world which may be entirely constructed out of language anyway. There is some doubt whether everybody reaches this final stage, which is probably just as well!

The mistakes children make can tell us a lot about how they are reasoning and where they are struggling. They show where there is missing information. In order to build up information they can use, children need to know *how to* think.

Encouragement and exploration, rather than parroting chunks of knowledge means that children get a chance to actually find a place for the new information that makes sense, fitting it in with what they already know. They *assimilate* it. When Homer Simpson gets a free flight as a bribe for not telling anyone about his illegal piloting experience he says:

"Because I endangered lives, we can fly anywhere we like!"
Homer Simpson: The Simpsons

When new information arrives that children don't understand, they make sense of it by providing a context for it to make sense in. This is called *accommodation*.

"Well I guess I learned my lesson – I lost creative control of the product and forgot to ask for money. Well, we live and learn."
Homer Simpson: The Simpsons

This oscillating push-me pull-you method of absorbing new information and building into and onto what we already know is the way we adapt to our environment and develop to make sense of our world. If we are forced by adverse circumstances to grow up too fast and make sense of too much, we bite off more than we can chew and generally make 4+4= fish, rather as Homer did in his assimilation above. It is frequently the task of the counsellor to unravel quite how we did it.

Though Piaget was the pioneer in this work, his theories have been modified and built on a great deal. They have been 'assimilated' and 'accommodated' into and onto greater knowledge as the years have gone by. This is very much the bare bones to give you a few working ideas. Reading lists for further exploration are included at the end of this appendix and the possibilities for refining and reframing and even disputing literally end with our own senility, but here is a bit more about *how* we think.

Assumptions:

When we make a wrong assumption or hold an inaccurate belief about something, it doesn't change the *way* we reach conclusions, although it will have a drastic impact on the conclusions that we reach. Rather like building a tower block on a swamp, once the foundations are in we will put breeze block on top of breeze block, insert steel rods to strengthen and apply cement to sandwich it all together. Unfortunately, the fact that it is on a swamp will eventually mean that the foundations shift, the building cracks and eventually the whole thing falls down. Like as not, someone will get hurt. What the cognitive psychologists (doctors of thinking) say about thought *processes* and the *way* that we learn is all true. Particularly in our formative years:

1. When we encounter a situation, we remember our response and what outcome our response achieved.

2. When we encounter similar situations, we notice whether varying our response makes a difference.

3. If it seems that nothing will make a difference, we stop making unnecessary responses.

These steps apply equally to learning that *no matter how much I feel, no one really sees me so I might as well stop feeling* as it does to, *if I scream and shout to get up because I don't want to go to bed yet no one listens so I might as well go to sleep.*

The second assumption here is appropriate for a young child at bedtime, but the first is not, especially if we become unconscious of what we really do believe.

Dogma:

Dogma is truth that has got stuck. In *Great Expectations*, the ancient Miss Havisham, 50 years in her wedding dress, her table strewn with the rat infested remnants of the wedding feast, has turned the truth of her broken, jilted heart into a dogma about men. The love and hopes she had about them have crystallised into despairing certainty like golden flowing honey in a jar becomes stale, solid sugar. She wreaks her revenge on men through Estella, who she has fed on a diet of isolation and hatred.

The worst thing about dogma is that it prevents people from seeing another way, or believing that there can be another way to see or to feel. Following the dramatic death of Miss Havisham when her wedding dress catches fire, Estella, far from being set free, follows what she knows to the point of *becoming* Miss Havisham, manifesting the reality inside her mind as if it were the only option. She shuts herself in with the rats and the cobwebby cake, keeps the curtains drawn and her heart closed. It takes true, unconditional love to break this spell (and it is a kind of magic). When Pip comes in he speaks the truth about Miss Havisham, drags the heavy dust laden curtains from the window and lets in the sun, which lights up the table like truth illuminating the reality of Miss Havisham's dreadful choices. A dogma insists that there is no choice; that there is only one reality.

Erikson

Erikson was a psychoanalyst and as such has no credibility here as a scientist. He had some good ideas though. His view of the unfolding person was that their successful development was dependent on managing to navigate conflicts and come out the other side stronger. Many of us have had the experience of standing up for what we believed, even when we were frightened, and discovering that we felt stronger for it. For Erikson the society of others generated conflict. He reckoned that in order to win through:

1. Trust had to overcome mistrust
2. Autonomy had to overcome doubt
3. Initiative had to overcome guilt
4. Industry had to overcome inferiority
5. Identity had to overcome role confusion
6. Intimacy had to overcome isolation
7. Creativity had to overcome stagnation
8. Ego integrity had to overcome despair

For Erikson you had to start at the top and work your way through the whole lot. This rather brings to mind the environmentalist in the Simpsons who refuses to take the coach to the save the tree demo, preferring instead to go in his

"Go-kart, powered my own sense of self satisfaction."
The Simpsons

According to Erikson, the struggle to overcome these conflicts made you stronger and more confident. To some it's a good reason to bring back national service, but of course things are never that simple. For instance, since I'm still at number one on the list I remain an emotional baby but, since I'm capable of working and writing this book, he can't have been all right. Just look at it as a few handy concepts for exploring things and finding out what the challenge is for today!

Since we're there, a brief dip into the ideas of Klein and Winnicot, which go by the alarming title of *Object Relations Theory*, seem worth a mention. These were also psychoanalysts. Klein's ideas I find absolutely terrifying. I still don't know whether what is terrifying about them is the product of her mind or mine, or maybe she's just right! Even thinking about it can really mess your head up so feel absolutely free to pass over to Winnicot, whose notion of the good enough mother is much more reassuring. You have permission to do this at any point. Since Klein herself underwent very traumatic experiences of bereavement, loss and separation, having to flee Hungary because it was so anti-semitic and was then ousted from Vienna because she was a radical woman, it seems likely that her own trauma became translated and truly felt, her theories penetrating depths of horrific emotional experiences none the less relevant for that. That doesn't mean however, that if you look it in the eye you won't go mad. Respect.

It is possible to emotionally overdose on these ideas. Do not operate machinery while thinking about them and handle yourself with care.

Melanie Klein:

"Therapy...breaks up families, turns wives against husbands, children against fathers, neighbours against me..."
Homer Simpson: The Simpsons

Klein theorised that the feelings of love and hate come about when we are babies. When we get what we need and long for, we experience love. When we have to wait and our pleasures are frustrated, we experience hate. Since the main caregiver is the one who both provides and falls short of providing this, then the main caregiver (then generally considered to be mother) becomes the person who the infant hurls both its love and its rage at.

Because the infant has not developed enough to get a solid sense of what is 'out there' (object permanence) it makes sense of its experience by imagining that the experiences of love and hate derive from two different people. A good mother and a bad mother. A fairy and a witch. Klein's view was that the baby imagined a good breast and a bad breast; separate sources for these very different experiences.

The bad breast/mother/witch is terrifying, being devoid of any love. The theory goes that the baby is so angry and filled with rage that in order not to self destruct, the infant pretends it is the bad breast/mother/witch who is so angry, rather than itself.

"She's gonna tell Marge to leave me. It'll
break up the family and you'll have to live
with your Grandma and pick beans."
(Homer Simpson talking about Marge's therapist)
The Simpsons

Then the baby can be justifiably filled with rage and have a focus because it is convinced it is being attacked by something worse. It is the only way the undeveloped small person can make sense of its experience. Klein called this the paranoid-schizoid position. Schizoid because of the way it splits its experience of reality into all-good and all-bad objects and paranoid because the '*something is coming to get you*' sensation arises from the baby but the baby thinks it's coming from outside. Scary stuff! Especially when, in *The Simpsons* we see Marge's psychiatrist say, reassuringly *"I'm not here to blame anyone*" as she surreptitiously writes *'Husband'* down on her notepad!

So that's supposedly standard for all of us. What happens then is that, as we get the hang of the fact that the mother is just one person (object permanence) we really start to freak out that with all our rage and hatred we have killed the very person we needed the most. You are now entering what Klein would call the depressive position. Filled with depressive anxiety and guilt, the child tries to make the mother feel better, repair the damage and make sure she is still alive. As long as she responds to the love given by giving love back, the child gets to see there's 'good and bad in everything'

So, a bit of *Ebony and Ivory* (*Stevie Wonder and Paul McCartney)* schmaltz and we all live happily ever after. All we ever needed to do was learn to integrate love and hate. Well, the ability to do that, as you might expect, seems to depend on just how bad things have been. Klein saw two ways that people avoided facing the loss of their illusions and having to learn the pain of compromise.

- They idealised the mother, fantasising that she was perfect. This meant that they split off the rage and directed it at some innocent bystander (a sort of 'whotchoolookinat' attitude)
- They manically denied that they loved her or needed her at all, so she just became a hate object that they could control through constantly fantasising about the worst thing she might do. In this way they'd be ready for it.

These defenses against feeling anxiety and guilt fall into two ways of relating. Idealising people and demonising people. The greater the pressure of these difficult feelings the more we split off from them. Recognising and, sometimes, holding these difficult and often ambivalent feelings without acting on them, is the task of the mature adult. When serious attachment crises occur in later life, like a divorce or the death of a parent, we are liable to fall back on this splitting as a way of coping.

D.W.Winnicott:

Marge: "An occasional hug is all I ask"
Bart: "Mum, you can hug me while I'm asleep"
Marge: "I do."
Lisa and Bart: "Eeegghh"
The Simpsons

Winnicott built on the work of Klein, in particular by studying what could make the child integrate all this grim reality! He reckoned that how we come to terms with all this difficult stuff is by having a good enough mother. He said the mother must:

- Hold the child. That is, not be too reactive to the child's behaviour, in order for it to feel safe and trust that the roof won't collapse if she/he expresses itself.
- Mirror the child in order to give it a sense of itself and what it wants:

"Go on Homer. Don't be nervous.
We all believe in you."(Marge)

- Optimally frustrate the child, in order for it to get a grip on the reality of who it is and what it can do while putting up with the frustration.

Without the 'holding environment' there would be no safe space for the child to really explore. For Winnicott this was about there being enough space, both psychic and physical, where the child could get a sense of itself as well as getting a sense of the limits and frustrations in life. It is this space that the counselling space, with its unconditional positive regard and its time boundaries seeks to emulate, and it is this space, in the 'not knowing' that becomes filled with fears and monsters and self hate if the formative experiences have been bad.

Without the safe space, without trust, the child develops a 'false self' which hides the 'true self' and so keeps it alive, but static, since it is not involved in developing through the give and take of everyday life:

'The good enough mother...starts off with an
almost complete adaptation to her infant's needs,
and as time proceeds she adapts less and less
completely, gradually, according to the infant's
growing ability to deal with her failure...'
Winnicott: Transitional Objects and Transitional
Phenomena 1951

In the case of the not good enough mother, what grows is the infant's false self as it becomes familiar with how to avoid, ignore and split off from her failure! These become defence mechanisms that we take forward with us into a dangerous world.

If the mother is excessive in reacting to the child by being hysterical when it cries or violent when it does something she doesn't like, then the child sees itself as threatened with annihilation. If the mother then soothes and apologises, the painful experience can be acknowledged and repaired, and the child's sense of itself becomes more constant. If the mother consistently attacks and fault is never acknowledged, then the child's personality becomes adapted to the expectation of constant attack. In my experience, even as adults, all that my clients ever really wanted was some acknowledgement of responsibility from the people who took advantage of their vulnerability.

Who am I? This is the constant unconscious question of the infant, who looks in its mother's face for the answer as if she were a mirror. The expression on the mother's face, the love or the hatred, the exhaustion or the liveliness is seen by the infant to be related to itself, to what she is looking at. It is from the mother's evaluation of the infant that the infant takes its cue and finds its answer.

Carl Rogers

Phheeww. I feel like I've reached a safe haven. Carl Rogers was the dad every little girl should have had. OK, there might be a bit of idealising going on here, but to admit that to old Carl would have been an excuse for a bit of a giggle rather than a deep delving into horror and that, basically, is what's great about him! He looked on the bright side without shirking responsibility for looking on the dark.

Carl Rogers was first and foremost a scientist with absolute respect for method, but he did point out a basic truth about science that many people miss. People are not method and method is not people; it is what people *do* with method that is important. For me, Carl Rogers is a cross between an entirely compassionate Dr. Who (light years ahead of everyone else) and the old professor in *The Lion, the Witch and the Wardrobe*. He is endlessly willing to admit that there's more going on than he could possibly know about. In this book about Narnia, Lucy insists to her older brothers and sister that she has been to another world inside a wardrobe. They check the wardrobe, and finding they can't get access they run to the professor, concerned that Lucy might be mad. He replies:

"There are only three possibilities. Either your sister is telling lies, or she is mad, or she is telling the truth. You know she doesn't tell lies and it is obvious she is not mad. For the moment then and unless any further evidence turns up, we must assume that she is telling the truth."
C.S.Lewis: The Lion, the Witch and the Wardrobe

This after muttering away to himself, rather endearingly:

"Logic...why don't they teach logic at these schools."
C.S.Lewis: The Lion, the Witch and the Wardrobe

What Carl Rogers questioned was the idea that science was something that could explain things with absolute certainty as if it already knew *all the* facts. In particular this was an issue when it studied people. As I have already said, the scientific method within psychology was always about studying behaviour from the outside, because it is observable. It cannot get 'inside' experience, for instance, to observe the feeling, but the problem with this approach is that treating people as objects is inevitable. Talking 'about' people only in terms of their behaviour ignores their wholly human aspect. In particular, it takes no notice of consciousness and the moral issues to do with being 'conscious'. Once this step has been taken it is easy to build up a huge amount of knowledge which can be used to manipulate people's behaviour.

"Where's Tipsy?' says Homer, looking forward to a friendly chat.

"Actually there is no Tipsy McStagger...
He's just a composite of other successful logos."
The Simpsons

The scientific method can predict which message in an advert is likely to make people buy more, but only 'people' can decide to run the adverts. The more knowledge we have about how others respond, the more scope we have for treating people like objects. Knowledge is power in the hands of often irrational, certainly biased and generally self-interested people, scientists or otherwise. Looked at in this way, the question that is more important than 'what do we know?' is 'what guidelines do we adopt for the use of what we know?' Ethics becomes of primary importance in the pursuit of knowledge.

Science, as we all know from Frankenstein's monster, has no ethics, no boundaries, and no appropriateness other than the rules we give it. Logic, too, is merely a tool; *a tool that helps us to explore what we think we know.* If we are rational we will, as the Professor above suggests, stop at what we don't know and await further information or *data* by which we can reassess our learning. Therefore science cannot tell us everything.

Answering the question 'does therapy work' immediately leads to another question. Work for who? Cut your hand off therapy stops people stealing quite effectively. Electric shock therapy has been known to turn people away from homosexuality. Chastity belt therapy does stop women from being promiscuous. What Rogers noticed was that science is always applied within a subjective, often cultural, value based context. It is done for non-objective *reasons.* It starts with some lunatic idea, like 'I wonder if the world is round' and ends up dragging everyone into a new way of seeing things *if* the people with the power aren't too threatened by this new way. (Copernicus was imprisoned)

Recognising this, it seemed to Rogers to be perfectly valid to study how subjective perception affected people's sense of well being. Rogers added to his method a fact that others missed: expert psychologists might not be able to get inside the experience of others but clients *are* conscious and can observe where no expert psychologist has observed before.

The mad idea that Rogers had, which he initially ignored, since it didn't fit too well with his sense of what he thought he knew, was that the best way to anticipate who would become a destructive teenager was to measure how realistic their *understanding of themselves* was as well as how much they *accepted* who they were. This he found to be true. As a psychologist, Rogers had assumed that family, environment, trauma, peers were all far more relevant. This new study changed his focus and really pulled it right out of the potentially pompous, ego-enabling arena of grand theories and sweeping judgements. He devoted the rest of his life to identifying measurable methods for improving the quality of people's lives and found that something very obvious was scientifically measurable and statistically true:

- That improvement came about via a relationship which had no exploitative aspects
- That because of this, his theory of therapy had an inbuilt way of optimizing the sharing of power and the empowerment of the client
- That the ethical considerations, if the method were followed, took care of themselves
- That this led from the client's self concern to the client's concern about the whole world

The simplest ideas are nearly always the most effective and generally the most threatening to those with the power. For instance, to empower others often means people like the big nurse in *One Flew over the Cuckoo's Nest* will have to give some up. Not for nothing was Jack Nicholson's character lobotomised.

The obvious question here is: 'how do we know that his type of therapy improved the quality of people's lives? How do you measure it?' The key is that Rogers didn't. The clients did. He set up the experiment and collated the data. Without going into too much detail here (and this research was very rigorous) Rogers and his cronies compared how a client's assessment of themselves changed in relation to a course of therapy. In particular, to what extent clients began to accept who they were. During therapy, feelings which have been previously denied come into awareness so that the person becomes more understanding, experiences less conflict, has a more tolerant attitude towards her/himself and a more tolerant, accepting attitude towards others. That's the theory.

Of course anyone could say anything. One of the checks was that, at the outset of the experiment a psychological profile or assessment was made by a clinician of all the people participating. At the end this was correlated with the self-assessments made.

The experiment compared those who had person centred therapy with those who didn't. The main research tool was a hundred statements about the self such as:

I am very relaxed
I am submissive
I am afraid of sex

At points throughout the year everyone had to sort these statement cards into nine piles ranging between 'most like me' to 'least like me'. They also had to sort them into piles 'most like I *would like to be*' and 'least like *I would like to be*'. The self we would like to be Rogers called 'the ideal self'. Because conflict between aspects of oneself and denial of aspects of oneself are partly the cause of difficult behaviour and inner suffering, *congruence* between them has to be an improvement. *Congruence* as we shall see in a moment, is one of Rogers's key ideas. In this context, when the ideal self and the real self are increasingly similar to one another, the person becomes increasingly *congruent.* An example of *incongruence* would be thinking one thing, feeling another and doing something at odds with either experience.

So, instead of asking 'Why are people disturbed?' He asked, 'What works to improve the congruence?' Of course, he had a hunch already and it was just a case of following his subjective idea with rigorous testing.

Rogers idea was that if you provided people in therapy with certain core conditions, they would become happier, better adjusted, less judgmental of themselves and others and more able to accept themselves: They *would value their whole selves more highly.* Those core conditions were:

- Unconditional positive regard (acceptance of them as a person)
- Empathy (understanding of their frame of reference)
- Congruence (on the part of the therapist)

And it's a lot more demanding than it looks because the therapist has to be genuine! Rogers was radical. Even subversive. For the first time the role of expert was turned on its head. The therapist became vulnerable. The feelings and awareness of the therapist became central and, for the therapist, hiding behind the 'expert' bunker of knowledge became anathema to helping the client 'become'. Not become cured or successful or even healthier. Just 'become' who s/he is. And trust that this will be a movement towards increasing health.

There is more information, but not much. There is more depth and understanding: too much for me to begin to put down here. Everyone has their own perceptions of what is important. Of what they need to express. Of how long they need to experience things for. There are, however, several obvious additions to the above:

- The client must be in psychological contact with the therapist – they must be able to connect.
- The client must be vulnerable, anxious or in need.
- The client must know or glimpse that the therapist accepts her/him as a person and has some empathy with her/him.

And that's it. So what does that remind you of? It seems to me that Rogers was the link between what the developmentalists knew was missing and the kind of chaotic and damaging behaviour we seem to be unable to rectify today.

What Rogers found was that the client became closer to the person they wished to become when they began therapy while accepting some less ideal aspects. This picture was significantly similar to the view the clinician had made of them at the outset. The client then, became more aware of their whole self and accepted it.

One of the most significant factors that is perhaps even *a necessary condition* in regard to this process is that the less the therapist is expert, the more s/he accepts how little s/he actually *knows,* the more able the client is to actually be who s/he is.

> *'the significant changes in the client have to do with very subtle and subjective experiences – inner choices, greater oneness within the whole person, a different feeling about oneself...a warmly human and genuine therapist, interested only in understanding the moment-by-moment*

feelings of this person who is coming
into being in the relationship with him,
is the most effective therapist.'
Carl Rogers: On Becoming a Person

What a radical idea. Think of the therapists and the doctors and the policemen and the teachers who hardly think of their clients as people at all. Think of how using power to hide behind a role or a sense of superiority prevents people from being who they truly are. If what Rogers discovered was true then massive resources that could go into helping people become healthy competent participating members of society are *actually* going into maintaining the inequality as the helping professions focus on what they think is important rather than what is really important.

How congruent! How symmetrical and pleasing and consistent. Oscar Wilde once said *'aesthetics are morals'*. I think that what he meant was that the method, the means, the manner of going forward has to fit with what we are trying to achieve. You cannot get people to be gentle and loving by force. You cannot get them to act morally by being immoral and you cannot get them to be vulnerable, and connect with their true, struggling, embittered or frightened self by being all-powerful and unable to connect with that vulnerable place in yourself.

'the individual must be willing to put his most passionate
beliefs and firm convictions to the impersonal test of
empirical research; but to be effective as a therapist, he
must use this knowledge only to enrich and enlarge his
subjective self, and must be that self, freely and without
fear, in his relationship to his client.'
Carl Rogers: On Becoming a Person

And how difficult is that? Sometimes, we therapists too have to settle for just being *good enough*! Perhaps acknowledging when we get it wrong is really at the heart of congruence, unconditional positive regard and acceptance!

Further Reading:

This is a very superficial smattering of the books available on this subject and some of them are original texts, often written for and by academics. I recommend that you look at these books and consider whether they are readable *for you.* If none of them suit, have a look for readers or more layperson friendly texts on www.amazon.co.uk or in good bookshops under *Developmental Psychology.* It's important that you buy something you feel able to digest.

Bowlby John: A Secure Base 1988 Routledge
Bowlby John: Attachment and Loss: Attachment 1997 Pimlico
Bowlby John: Attachment and Loss: Separation – Anxiety and Anger 1998 Pimlico
Bowlby John: Attachment and Loss: Loss – Sadness and Depression 1991 Penguin
Erikson E.H.: Identity and the Life Cycle 1980 W.W.Norton
Goleman Daniel: Emotional Intelligence 1996 Bloomsbury
Kirchenbaum H. and Land Henderson V. (eds.): The Carl Rogers Reader 1990
Constable and Robinson
Miller Alice: The Drama of Being a Child 1995 Virago
Mitchell Juliet (ed.) The Selected Melanie Klein 1987 The Free Press
Rogers Carl: Personal Power: Inner Strength and it's Revolutionary Impact 1978 Constable and Robinson
Winnicott D.W.: Home Is Where We Start From 1990 Penguin
Winnicott D.W.: The Family and Individual Development 1968 Routledge

Acknowledgements

Usually acknowledgements go to close family members – the people who have put up with you and helped you by reading your book and I am really very grateful to Peter Rawlings for his endless encouragement, to Gwen Newman and Gill Pirie for their patience and feedback, Tony Keele for creative cover making and tremendous art work and to my husband Oliver Perry for his editing skills.

Most of all though I am utterly indebted to the texts quoted in this book, all of which I continue to grown up with. My grateful thanks to all those involved in the making of the films: *The Sound Of Music; Trainspotting; Videodrome; Beyond Rangoon* and *The Killing Fields* and also to the writers and programme makers of *The Simpsons, Buffy The Vampire Slayer* and those responsible for the Slater family in *Eastenders.*

Lyrics from songs quoted include *What Is True* by the late great *Sandy Denny; Don't Think Twice, It's Alright* by *Bob Dylan*; *Rise* by *Public Image Limited*; *I don't Like Mondays* by *The Boomtown Rats* and *Born Slippy* by *Underworld.* My thanks also to the bands *Arrested Development* and *Rage Against The Machine* for adopting such great names.

Brief but nonetheless important acknowledgement goes to *Alice Miller* for *For Your Own Good*; *Louise Hay* for *You can Heal Your Life; Miller Mair* for *Between Psychology and Psychotherapy; Starhawk* for *The Spiral Dance* and to *Irving Yalom* for his notion of *Cyclotherapy.*

I have *Jamie Sams* to thank for her reframing of the word *responsibility* to *respondability, Philip Larkin* for his insight into the function of parents, *Nietzche* for his optimism, *J.M.Barrie* for *Peter Pan*, *T.S.Eliot* for plumbing unspeakable depths so articulately in his poem *The Four Quartets* and *Rabindranath Tagore* for his beautiful *Gitanjali* poems. Thanks also for the fairy tales of *Hans Anderson* and many others whose authors lie unnamed within the oral tradition of folk stories: *Snow White; The Snow Queen; Pinocchio; Sleeping Beauty; The Little Mermaid* and *The Bluebird of Happiness.*

Catch 22 by *Joseph Heller, George Orwell's 1984 and Ken Kesey's One Flew Over The Cuckoo's Nest* also hold a crucial part in this book, though barely mentioned but enormous gratitude goes towards those modern novels that I have quoted extensively and that provided the original inspiration. They are *The Lord Of The Rings* by *J.R.R.Tolkien;* The Harry Potter books by *J.K.Rowling*, in particular *Harry Potter and the Prisoner of Azkaban* and *Harry Potter and the Goblet of Fire; The Subtle Knife* and *The Amber Spyglass,* part of the trilogy *His Dark Materials* by *Philip Pullman; Slaves of the Mastery* and *Firesong* which are part of *William Nicholson's* trilogy *The Wind On Fire.*

Last I have to thank *C.S.Lewis* for his *Narnia* books. Particularly *The Lion, the Witch and the Wardrobe, The Magician's Nephew, Prince Caspian, The Voyage of the Dawn Treader, The Silver Chair* and finally, *The Last Battle* for bringing me home.

References

Films:

Beyond Rangoon: Columbia/Tri-star videos *d.w.*John Boorman
The Sound Of Music: TCF/Argyle *d*Robert Wise *w*Ernest Lehman
Trainspotting: Universal Pictures video *d*Danny Boyle *w*Irvine Welch
Videodrome: Filmplan International *d.w.*David Cronenberg

Television Programmes:

Buffy The Vampire Slayer: 20th Century Fox *w*Joss Whedon
Eastenders: BBC Worldwide Publishing *w*Tony Jordan
The Simpsons: 20th Century Fox *d*Matthew Nastik/Swinton O. Scott

Songs:

Denny S.: What Is True? Jardiniere Music
Dylan Bob: Don't Think Twice, It's Alright Columbia USA
Public Image Limited: Rise EMI Int'l
Sondheim Steven: Officer Krupke
The Boomtown Rats: I don't Like Mondays Vertigo

Non-fiction:

Hay L.: *You can Heal Your Life* *2003* Full Circle
Karpinski G.: *Where Two Worlds Touch* *1990* Ballantine
Mair Miller: *Between Psychology and Psychotherapy* *1989* Routledge
Miller A.: *For Your Own Good* 1987 Virago
Rogers C.: *On Becoming A Person* 1961 Constable
Starhawk: *The Spiral Dance* 1999 HarperSanFrancisco
Winnicott: *Transitional Objects And Transitional Phenomena* *1951* Heinemann

Poetry:

Eliot T.S.: *Four Quartets* 1944 Faber and Faber
Larkin P: *This Be The Verse* *Collected Poems* 2003 Faber and Faber
Tagore Rabindranath: *Gitanjali* 2000 Dover Publications

Fiction:

Barrie J.M.: *Peter Pan* 1993 Wordsworth Classics
Heller Joseph: *Catch 22* 1962 Jonathan Cape
Kesey K: *One Flew Over The Cuckoo's Nest* 1962 Viking
Lewis C.S.: The *Lion, the Witch and the Wardrobe* 1950 Geoffrey Bles.
Prince Caspian 1951 Geoffrey Bles.
The Voyage of the Dawn Treader 1952 Geoffrey Bles.
The Silver Chair *1953* Geoffrey Bles.
The Last Battle 1955 Bodley Head
The Magician's Nephew *1956* Bodley Head
Nicholson William: from *The Wind On Fire Trilogy:*
Slaves Of The Mastery 2001 Egmont
Firesong *2002* Egmont
Orwell G.: 1984 1949 Penguin
Pullman Philip: *His Dark Materials:* *The Amber Spyglass*
2000 Scholastic
Rowling J.K.: *Harry Potter and the Prisoner of Azkaban*
2000 Bloomsbury
Harry Potter and the Goblet of Fire 2001 Bloomsbury
Tolkien J.R.R.: *The Lord Of The Rings* *1954/5* HarperCollins

The author and publishers gratefully acknowledge permission to quote from the following works:

EASTENDERS (broadcast October 2001) by Tony Jordan, BBC Worldwide Publishing.